Stand-Alone Solar Electric Systems

Stand-Alone Solar Electric Systems

The Earthscan Expert Handbook for Planning, Design and Installation

Mark Hankins

SERIES EDITOR:
FRANK JACKSON

publishing for a sustainable future
London • Washington, DC

First published in 2010 by Earthscan

This book draws on some material that originally appeared in the author's book *Solar Electric Systems for Africa,* published by AGROTEC and Commonwealth Science Council in June 1995.

While the author and the publishers believe that the information and guidance given in this work are correct, all parties must rely upon their own skill and judgement when making use of them – it is not meant to be a replacement for manufacturer's instructions and legal technical codes. Neither the author nor the publisher assumes any liability for any loss or damage caused by any error or omission in the work. Any and all such liability is disclaimed.

This book was written using principally metric units. However, for ease of reference by readers more familiar with imperial units, the publisher has inserted these in the text in brackets after their metric equivalents. Please note that some conversions may have been rounded up or down for the purposes of clarity.

Earthscan Ltd, Dunstan House, 14a St Cross Street, London EC1N 8XA, UK
Earthscan LLC, 1616 P Street, NW, Washington, DC 20036, USA

Earthscan publishes in association with the International Institute for Environment and Development

For more information on Earthscan publications, see www.earthscan.co.uk or write to earthinfo@earthscan.co.uk

ISBN 978-1-84407-713-7

Typeset by Domex e-Data, India
Cover design by Yvonne Booth

A catalogue record for this book is available from the British Library

Library of Congress Cataloging-in-Publication Data

Hankins, Mark.
Stand-alone solar electric systems : the Earthscan expert handbook for planning, design, and installation / Mark Hankins.
p. cm.
Includes bibliographical references and index.
ISBN 978-1-84407-713-7 (hardback)
1. Photovoltaic power systems. I. Title.
TK1087.H36 2010
621.31'244--dc22
2009051707

At Earthscan we strive to minimize our environmental impacts and carbon footprint through reducing waste, recycling and offsetting our CO_2 emissions, including those created through publication of this book. For more details of our environmental policy, see www.earthscan.co.uk.

Printed and bound in the UK by Scotprint, an ISO 14001 accredited company. The paper used is FSC certified and the inks are vegetable based.

Contents

List of Figures, Tables and Boxes

Figures

Tables

Boxes

Preface and Acknowledgements

This book is for people who want to learn how to design and install small off-grid solar PV systems. It incorporates much content from a book that I wrote in the mid-nineties, *Solar Electric Systems for Africa*, which was supported by the Commonwealth Science Council in the UK. This new book has been fully rewritten, and has been greatly expanded based on rapid advances in solar, battery, charge controller, inverter and lighting technology.

Off-grid solar electricity is a way of life. It is gratifying to be able to switch on solar-powered lights, laptop, music or internet in a wild remote location, far from power lines and diesel generators. This smugness is even more complete when one has fixed the array, laid the cables, filled the batteries and bolted the charge controller to the wall oneself. That said, be warned that those bitten by the 'solar bug' are likely to become lifelong solar exponents and practitioners. So good luck, be careful and have fun!

Thanks are due to a number of people who helped me put this book together: Frank Jackson, the technical editor, kept me on course and served as teacher, resource person and photographer throughout. Thanks to Michael Fell, Claire Lamont, Hamish Ironside and the rest of the folks at Earthscan for the patience and professionalism to take this from manuscript to manual. Finally, thanks to my wife Gladys and my daughter Ayanna for their full support to this project.

This book is dedicated to three off-grid PV practitioners – true 'solar lions' who did much to spread PV technology around the world. Each of them taught many about the nuts and bolts of off-grid solar and helped lay the groundwork for this book. They will be remembered for the part they played in building a solar future in developing countries.

Walt Ratterman travelled to the most inhospitable, remote parts of the world to install off-grid PV systems and teach solar. An accomplished electrician and successful contractor, he was as comfortable lecturing about 50Wp systems as he was 50kWp systems. Though he got into solar rather late in life, he brought the inquisitive nature, cheer and dedication of a 20-year-old to the cause. For a short time, we worked closely together in Rwanda – and during the writing of this book he sent comments on first drafts to me from Lesotho, Haiti, Oregon and Palestine. While on a solar installation in Haiti, he died when his hotel collapsed in the Port-au-Prince earthquake of 12 January 2010.

Gaspar Makale, a capable solar technician, businessman and trainer, helped me and Frank Jackson set up the KARADEA Solar Training Facility in western Tanzania. Between 1993 and 2001, in the isolated hills of Karagwe, our small team trained hundreds of solar technicians from all over Africa. Working with Gaspar on projects in Kenya, Tanzania, Malawi and Uganda was always gratifying because – in harsh conditions with too few tools and too little money – he always had a surplus of good humour and enough cash to buy everyone rounds at the bush pub when the job was done. He died at home in Kagera, Tanzania in 2007.

Harry Burris inspired me to get involved with PV in 1983. More than any single person, Harry was responsible for the early growth of solar electricity markets in Kenya, Zimbabwe and Tanzania. He introduced the concept of the solar home system to East Africa: building supply chains, spreading the word to solar companies, convincing Kenya's battery manufacturer to introduce a 'solar battery' and working with local electronics companies to assemble 12V DC lights. Travelling by bicycle and bus, staying in dodgy hotels, and living like a local, Harry and his team installed thousands of systems in remote homes, schools and clinics over 5 years, and he trained dozens of Kenya's first technicians in the process. During 1995 in Zimbabwe, he was leader of the UN's first solar project in Africa, and later, in Tanzania, he was known as 'Baba Solar' in the last years of his life. In 1985, on his World War II correspondent's typewriter, together we drafted the training material that eventually became the core of this book.

Mark Hankins
Nairobi, April 2010

List of Abbreviations

AC	alternating current
AGM	absorbed glass mat
BoS	balance of systems
CdTe	cadmium telluride
CIGS	copper-indium-gallium-diselenide
CSP	concentrated solar power
DC	direct current
DoD	depth of discharge
HIT	heterojunction within intrinsic thin layer
HVD	high-voltage disconnect
Isc	short-circuit current
LED	light-emitting diode
LPG	liquefied petroleum gas
LVD	low-voltage disconnect
MCB	miniature circuit breaker
MPPT	maximum power point tracker
nicad	nickel cadmium
NOCT	normal operating cell temperature
PSH	peak sun hours
PV	photovoltaic
PWM	pulse-width modulation
SHS	solar home systems
SLI	starting, lighting and ignition
SoC	state of charge
SWH	solar water-heater
STC	Standard Test Conditions
UPS	uninterruptible power supply
Voc	open circuit voltage
VRLA	valve-regulated lead-acid batteries
Wp	watt-peak

Note: the terms 'solar electric', 'solar PV' and 'PV' mean the same thing and are used interchangeably throughout the text.

1

How to Use this Book: An Overview of Solar Electric Technology

Off-grid solar electricity is a convenient form of electricity for people far from the electric grid or for people who want electric power without having to 'hook up'. Whether the requirement is a cabin in the woods, a rural house in a developing country, a cruising sailboat or a garden shed, this book can help you to plan and install a solar system so that it robustly meets the required needs. Moreover, with this book you will also be able to understand the electricity flows in the system you use – a critical concept in a world increasingly short on energy. Most of all, you will be able to proudly say that you harvest your own power – and you did it yourself!

This chapter gives readers a basic understanding of the common components of off-grid solar electric systems. It also provides a brief background of solar electric technology, summarizing the applications, parts, advantages and disadvantages of solar electric systems including solar modules, batteries, charge controllers, inverters and appliances. The end of the chapter has a basic introduction to common electric terminology.

Common Uses of Solar Electricity

Solar electricity is electric power generated from sunlight using devices called solar cell modules. Solar electricity can replace, cost-effectively, small applications of petroleum-fuelled generators, grid power and even dry cell batteries. The technology has spread rapidly throughout the world for both on-grid and off-grid application. Millions of rural off-grid homes are using solar photovoltaic (PV) systems throughout the developed and developing world.

Small off-grid solar electric systems differ from grid or generator electricity in a number of ways:

- Small off-grid PV systems are based on extra-low-voltage direct current electricity, not low-voltage 230 or 110 volts (V) alternating current (see the Glossary for a definition of 'extra-low' and 'low' voltage).
- Off-grid PV systems usually store energy in batteries.
- Electricity is generated on-site by photovoltaic modules.
- For systems to be economical, all electricity produced must be used efficiently.

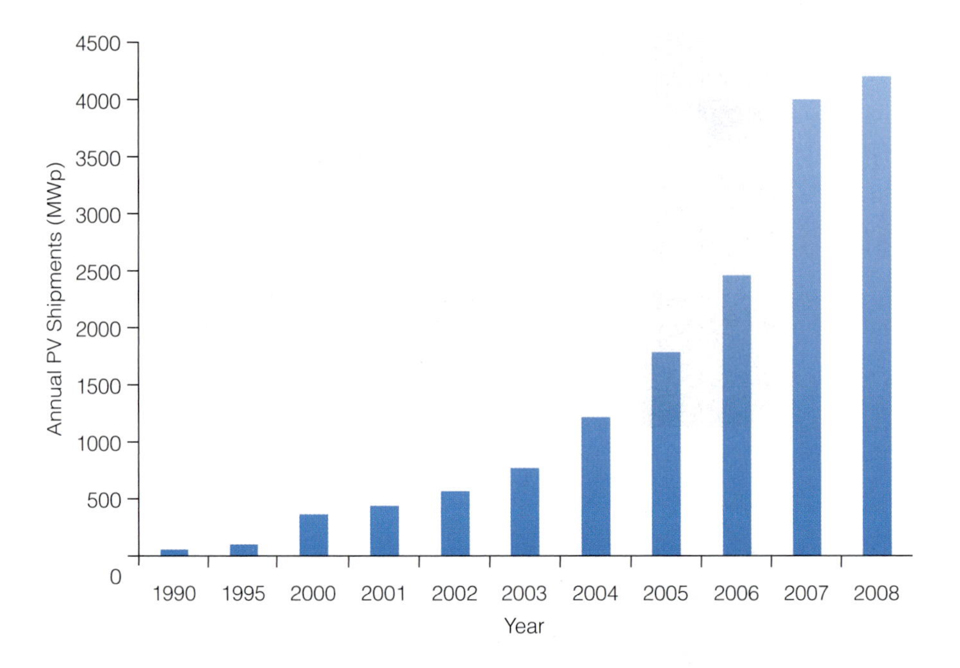

Figure 1.1 *World shipments of PV modules, 1990–2008*

Solar electric technology is developing very quickly; its worldwide use is increasing rapidly as prices of other electric energy sources rise. Before the 1990s, solar electricity was a new technology, used mostly for off-grid telecommunications systems, signalling, water pumping and remote clinic power.

Since the mid-1990s, world production of solar modules increased rapidly, largely because of demand for grid-connected systems in Europe, the US and Japan. Meanwhile, the price of solar cell modules fell from about US$100 per peak watt in 1974 to less than US$4 per peak watt in 2008. Decreased prices caused a rapid expansion of solar PV into rural markets in the early 1990s. Today, solar electricity is often the most viable power source for many off-grid applications in rural areas that have not been electrified.

Off-grid and grid-connected solar PV systems are very different, both in their component make-up and in the principles used to design them. Off-grid solar (also called 'stand-alone') systems usually require battery banks to store energy and they must be carefully designed to meet expected loads during cloudy periods. Meanwhile, grid-connected systems feed unused power directly into the grid and usually do not utilize battery storage systems.

This book is a general guide for installers of small off-grid systems below 1 kilowatt peak in size to be used in conjunction with system component installation manuals and national electrical codes and regulations, which should always be referred to and complied with. Chapter 11 deals specifically with installing systems of over 500Wp. Lists of literature on installing larger systems and grid-tied systems can be found in Chapter 12 (as well as from providers of training courses). Installation work should always be carried out by (or with) appropriately qualified technicians.

(a) *Solar lighting system on a girls' boarding school in Tanzania*

(b) *Pole-mounted panels at a health centre in Rwanda (Walt Ratterman is second from right)*

(c) *PV system in a Kenyan game park*

Figure 1.2 *Common PV applications*

Sources: (a) Frank Jackson; (b, c) Mark Hankins

(d) *Ten year old solar water pumping system in Puntland, Somalia*

(e) *Small lighting system in Bangladesh*

Figure 1.2 *(Continued)*

Sources: (d) Mark Hankins; (e) MicroEnergy International

The list below describes some of the important applications for which off-grid solar electric power is utilized (see Chapter 11 for case studies on different types of systems).

Household lights, televisions, sound systems, radios and small appliances

Often known as 'solar home systems' (SHS), small PV systems provide electricity for lighting and entertainment appliances. Night-time lighting is crucial for education, commerce, craft work and all types of social activity. Off-grid rural people need information and entertainment, and they value televisions, stereo systems, computers and communication devices.

Small industries and institutions

Schools and small businesses in rural areas use solar electricity to power lights, small machines, calculators, light tools, computers, typewriters, communication devices and security systems.

Telecommunications

Because telecommunication systems are often installed in isolated places with no access to power, stand-alone photovoltaic systems are a common choice to power radios, remote repeaters, base stations for cell-phone networks and weather-monitoring equipment.

Health centre vaccine refrigeration and lighting

Solar electric systems are popular for vaccine refrigeration in rural health centres. Such solar refrigerators are also utilized to freeze ice packs and to keep blood plasma cool. The World Health Organization supports programmes that install solar electric refrigerators and lighting in health centres around the world.

Water pumping

Arrays of solar cell modules connected to electric pumps are used to pump water from wells or boreholes. This water is used for drinking, washing, other household purposes and for small irrigation projects. Solar electricity can also be used to help purify drinking water. It is relatively expensive to use solar PV for commercial-scale irrigation.

Electric fencing and other uses

Solar electric fences keep wild animals inside game parks (and out of farm land), while also keeping domestic animals within designated areas. Other common uses of solar electric systems include street-lighting, road-sign illumination, railway and marine signal lighting, security systems and protection of pipelines from corrosion.

Table 1.1 *Solar electricity advantages and disadvantages*

Advantages of solar electric power	Disadvantages of solar electric power
• PV systems consume no fuel and convert freely available sunlight directly into electricity. • PV systems produce electricity quietly without giving off exhaust gases or pollutants. • PV systems require comparatively little maintenance. Solar modules have no moving parts and last over 20 years. • PV systems are particularly economical for small applications. Applications where power demand is below 3–5kWh/day are particularly cost effective using solar PV. • PV systems can be tailored to the size of the application needed, be it lighting, pumping or audio-visual, and they can be easily expanded as demand increases. • PV systems are safe when properly installed. Risks of electric shock are low with 12 and 24 volt DC systems, and there is much less fire-risk than for kerosene or generator solutions.	• PV systems often have higher up-front costs than other solutions. Even if a solar PV system cost is more economical than generators or kerosene over its lifetime, it is often difficult for low-income people to access cash to buy the system up-front. • Most off-grid PV systems require batteries to store electric power. Batteries require maintenance and must be replaced at the end of their lives. The performance of PV systems is dependent on the quality of batteries available on the local market or the availability of imported batteries. • Small PV systems often require efficient or direct current (DC) appliances. These often cost more than commonly available alternating current (AC) appliances. • PV systems must be designed and installed by qualified technicians. Poorly designed or installed PV systems perform less effectively than other solutions. • Large stand-alone PV systems often need to be backed up by petroleum-fuelled generators (or wind-power systems) to supply power during peak-use or cloudy periods. • Solar electric systems are not economical for thermal loads such as cooking, water heating or ironing clothes.

Advantages and Disadvantages of Solar Electric (PV) Off-Grid Systems

When deciding on a system, always consider the advantages and disadvantages of the technology for your particular requirements. The table above summarizes some of these for solar PV systems.

Using this Book to Design and Install Solar Electric Systems

The following section guides the reader through the parts of a solar electric system, noting which chapters have detailed information about each topic.

Fundamentals of solar energy: how much energy is available and how it can be used (Chapter 2)

When planning a solar electric system, first ask whether your location has enough solar energy available. The output of a solar module depends on the amount of sunlight falling on it, so it is important to plan a PV system with weather patterns in mind. In most areas, solar energy is plentiful, although at certain times of the year there is often less energy. Chapter 2 also contains basic information about solar energy principles and solar devices.

Solar cell modules: harvesting solar energy (Chapter 3)

Solar cell modules are devices used to convert sunlight into electricity. Because there are many types of solar modules, the module required for a given task should be chosen carefully. Modules should be mounted so that they can collect maximum energy.

Batteries: storing solar energy (Chapter 4)

Appliances are often used after sunset – when solar modules are not producing electricity. Batteries store energy collected during sunny days for use at night and on cloudy days. Various types and sizes of batteries are available. The type of battery chosen depends on the energy requirements of a system and the budget of the user. Batteries need to be maintained and monitored to ensure that they have a long life.

Charge controllers and inverters: managing solar energy (Chapter 5)

Solar cell modules supply a limited amount of energy. For this reason, solar electric systems must be managed so that the energy collected by the solar cell modules balances the amount of electricity used to power lights and appliances. Charge controllers are used to prevent damage to batteries and other parts of the system from overcharging and deep discharging. They may also alert the user when the battery or module is not functioning properly. Inverters convert direct current electricity from batteries into alternating current electricity used by many appliances. It is important to select the right type of controller and inverter for your needs.

Lamps and appliances: using solar energy efficiently (Chapter 6)

Unlike generators or mains (which supply AC electricity at 240 or 110 volts), solar electric systems operate on extra-low-voltage direct current. For this reason, special DC lamps and appliances are often installed in PV systems. Furthermore, efficient fluorescent or light-emitting diode (LED) lamps are usually used in solar electric systems because they consume far less power than incandescent-type lamps. Choose lighting devices, fixtures and appliances carefully with PV systems.

Wiring and fittings for solar electric systems (Chapter 7)

Wiring of solar electric systems is similar to the wiring of 'grid-based' electricity systems. However, extra-low-voltage solar electric systems usually require thicker wire than grid systems and some fittings differ from those used in 240V or 110V AC systems. When designing PV systems, planners should carefully consider voltage drops and select the correct wire sizes and fittings.

Designing and planning PV systems (Chapter 8)

In order to properly plan a solar electric system, you must:

- determine how much energy is needed to power your lamps and appliances by adding up the number of lamps and appliances, and tabulating their power demand and their daily usage time;
- estimate how much solar energy is available at the site per day;
- estimate how much energy is lost due to inefficiencies in the system;
- determine the size of array, control and battery required;
- make a plan for procuring all of the needed components and fittings.

Installing PV systems (Chapter 9)

If a solar electric system is planned properly, installing the components should proceed smoothly. Nevertheless, there are procedures that should be followed during system installation and there are electrical standards that need to be adhered to for safety reasons and to ensure that the equipment is not damaged. Remember, because off-grid solar installations are often conducted in remote areas, it is critical to bring all necessary tools and components to the site.

Maintenance and service practices (Chapter 10)

Once a system is installed, it must be serviced on a regular basis to ensure proper operation. Tasks include regular maintenance of batteries and circuits, and cleaning of the solar cell modules. Owners and operators must know how to operate and maintain all the parts of their PV system.

Specialized and large off-grid PV systems (Chapter 11)

This chapter is about the special needs of larger, specialized or hybrid PV systems. It contains annotated case studies of solar PV systems, including clinic, school, pumping and larger residential systems, that may help designers plan their own systems. It also has information about appliances used in large systems.

Resources for solar planners and technicians (Chapter 12)

This chapter contains information about additional information sources for practitioners and interested consumers. Information includes websites, books and publications, organizations and sources of useful codes of practice and standards that have been developed.

Box 1.1 Basic extra-low-voltage electric concepts

You need a basic knowledge of electricity to install and design solar electric systems. The table below defines some of the common terms used in this book. Once you have mastered the terms and principles in the table below, then the following chapters should be readily understandable. Those who are used to working with 230V AC systems will find that extra-low-voltage DC systems are safer (from the point of view of shock hazard) and easier to work with.

For anyone who has not had experience with electricity, Appendix 1 explains electrical concepts in more detail, and goes through several examples. Always seek the advice of a qualified electrician or solar technician when installing a system if you are not familiar with electricity.

Table 1.2 *A summary of basic electric concepts*

Term	Symbol	Unit (Abbreviation)	Definition
Current	I	amps (A)	Rate of flow of electrons through a circuit.
Direct current	DC	amps (A)	Flow of electric charge which does not change direction with time. Solar cells always produce direct current.
Alternating current	AC	amps (A)	Electric current which first flows through the wire in one direction and then the other, and continues to switch back and forth over time.
Voltage or potential difference	V	volts (V)	The difference in potential energy between the ends of a conductor that governs the rate of flow of current.
Resistance	R	ohms (Ω)	The property of a conductor (i.e. a wire or appliance) which opposes the flow of current through it and converts electrical energy into heat.
Electric power	P	watts (W)	Rate at which energy is supplied from the power source or consumed by an appliance.

A 'circuit' is a system of conductors (i.e. wires and appliances) capable of providing a closed path for electrons. Current can flow when the circuit is closed. No current can flow when the circuit is open.

The 'load' is the set of equipment or appliances that uses the electrical power from the generating source, battery or module. Series and parallel refers to the arrangement of the load and batteries within the circuit (see Appendix 2).

The 'Power Law' states that electric power (in watts) is equal to the voltage (V, in volts) multiplied by the current (I, in amps).

$$\text{Power (watts)} = \text{Voltage (volts)} \times \text{Current (amps)}$$

'Ohm's Law' states that the voltage of a circuit is equal to the current (in amps) times the resistance (in ohms). This law is particularly useful in determining how much voltage is lost on long wire runs.

$$\text{Voltage (volts)} = \text{Current (amps)} \times \text{Resistance (ohms)}$$

2

Fundamentals of Solar Energy

This chapter explains the basic principles of solar energy. Diffuse and direct solar radiation, irradiance and insolation (the quantity of solar radiation falling upon an area) are discussed. Estimating energy available at a given site using meteorological records is outlined and the use of tracking to increase energy system output is also examined. The final section describes basic principles of collecting solar energy including reflection, concentration and the 'greenhouse effect'. Common non-electric uses of solar energy are outlined, including crop-drying, water-heating and solar-cooking.

The Solar Resource

The sun is the Earth's nearest star and the source of virtually all the Earth's energy, producing 3.8×10^{23} kW of power in huge nuclear fission reactions. Most of this power is lost in space, but the tiny fraction that does reach the Earth, 1.73×10^{16} kW, is thousands of times more than enough to provide all of humanity's energy needs.

The energy we derive from wood fuel, petroleum products, coal, hydroelectricity and even our food originates indirectly from the sun. Solar energy is captured and stored by plants. We use this energy when we burn firewood or eat food. The sun also powers rainfall cycles that fill rivers from which we extract hydroelectricity. Petroleum and coal are made up of the fossilized remains of plants and animals that collected energy from the sun thousands of years ago.

This book is concerned mostly with off-grid use of solar electricity – the use of solar cells (also called photovoltaic or, for short, PV devices) to harvest electricity from the sun. Chapters 2 to 12 go into detail on solar electric technologies and applications.

Energy can also be harvested from the sun directly for heating, drying, cooking, distilling, raising steam and generating electricity. Many types of equipment can be used to collect solar energy. These include flat plate solar-thermal panels and evacuated tubes, which harvest solar energy for heating water, and solar concentrators that focus the rays of the sun into high energy beams to produce heat for electricity generation (known as concentrated solar power or CSP).

Historically, collecting and harnessing solar power has not been as easy or convenient as it has been for other energy sources, for several reasons. First, energy from the sun is spread over a wide area in a relatively low energy form. Unlike petrol or coal, which are high-energy and can be easily transported, solar energy arrives in a scattered manner that is difficult to usefully trap, convert and store. In order to collect it, solar energy harnessing equipment must be utilized. Secondly, solar energy is not available at night or during overcast and cloudy weather, and the forms of energy derived from solar energy must be stored. This means additional equipment must often be used to store the energy; this accounts for a large portion of the costs of solar energy systems.

In the past, solar energy has often been overlooked because of the high price of the equipment used to harvest and store it. However, as the prices of other energy sources such as petroleum fuel, biomass and even coal-generated electricity rise – and as the environmental risks associated with other power sources are increasingly recognized – solar energy equipment is fast becoming economically attractive.

Converting Solar Energy

Solar energy is plentiful worldwide. Most people do not stop to think about how solar energy heats their homes or provides energy to grow crops. Sunshine has traditionally been used for drying all types of things: clothes, agricultural produce, cash crops and bricks – even in the production of salt from sea water.

To make use of solar energy, we must convert it into useful forms. As shown in Figure 2.1, solar energy can be usefully transformed in three ways:

- solar energy to chemical energy;
- solar energy to heat energy;
- solar energy to electrical energy.

Solar energy to chemical energy

Green plants transform solar energy to chemical energy in sugar and cellulose by the process of photosynthesis (all biomass contains chemically stored solar energy). Unfortunately, we have not yet developed a way to directly transform solar energy into chemical energy. Photosynthesis remains a secret of plants!

Solar energy to heat energy

Solar heating devices transform solar energy into heat that is used for drying, water-heating, space-heating, cooking and distilling water. CSP plants convert water to steam that is used to generate electricity.

Solar thermal energy is most easily used in applications that require relatively small amounts of heat. The cheapest and simplest uses of solar energy (e.g. solar driers and water heaters) raise the temperature of air or water by 20–40°C (36–72°F). When more energy and higher temperatures are needed, solar energy must be concentrated, transported and/or stored, greatly increasing the cost and complexity of solar equipment needed.

Solar energy to electrical energy

Solar electric devices transform solar energy into electrical energy. This can be used to directly power electrical devices such as pumps and fans – or it

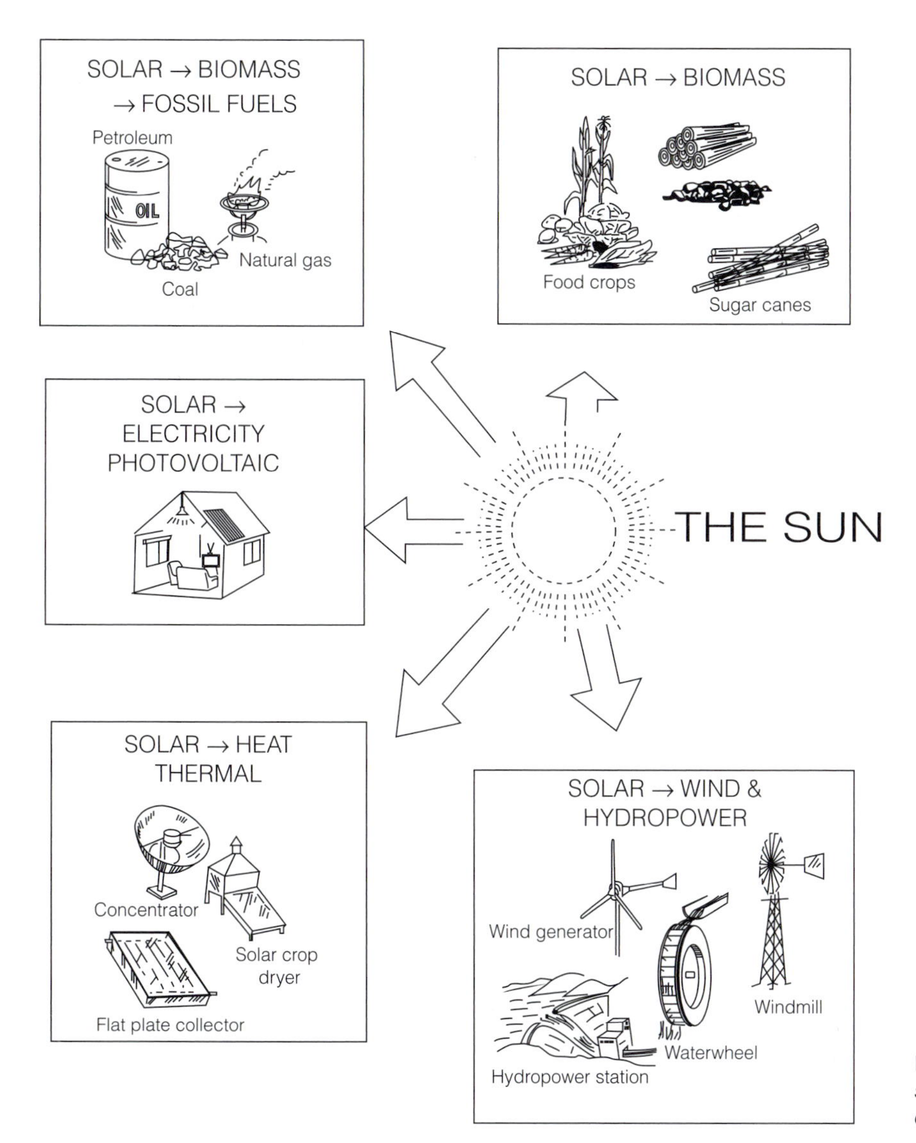

Figure 2.1 *The sun is the source of virtually all of our energy*

can be stored in batteries to power lights, televisions, refrigerators and other appliances (these appliances are often used mainly at night when the sun has gone down).

Solar Radiation Principles

Sunshine reaches the earth as a type of energy called radiation. Radiation is composed of millions of high-energy particles called photons. Each unit of solar radiation, or photon, carries a fixed amount of energy. Depending on the amount of energy that it carries, solar radiation falls into different categories including infrared (i.e. heat), visible (radiation that we can see) and ultraviolet (very high energy radiation). The solar spectrum describes all of these groups

Box 2.1 Basic energy and power concepts

Energy

Energy is referred to as the ability to do work. Energy is measured in units called joules (J) or in watt-hours as shown below. One kilojoule (kJ) is equal to 1000 joules and 1 megajoule (MJ) is equal to 1 million joules.

Watt-hours (Wh) are a convenient way of measuring electrical energy. One watt-hour is equal to a constant 1 watt supply of power supplied over 1 hour (3600 seconds). If a light-bulb is rated at 40 watts, in 1 hour it will use 40Wh, and in 6 hours it will use 240Wh of energy. Electric power companies measure the amount of energy supplied to customers in kilowatt-hours (kWh) or thousands of watt-hours. One kilowatt-hour is equal to 3.6 megajoules.

Energy conversions

watt-hours × 1000 = kilowatt-hours

kilowatt-hours × 1000 = megawatt-hours

megajoules ÷ 3.6 = kilowatt-hours (or peak sun hours)

kilowatt-hours × 3.6 = megajoules

Power

Power is the rate at which energy is supplied (or energy per unit time). Power is measured in watts. One watt is equal to 1 joule supplied per second.

Power conversions

watts ÷ 746 = horsepower

watts × 1000 = kilowatt

kilowatts × 1000 = megawatts

of radiation energy that are constantly arriving from the sun, and categorizes them according to their wavelength. Different solar cells and solar energy-collecting devices make use of different parts of the solar spectrum.

Solar energy arrives at the edge of the Earth's atmosphere at a constant rate of about 1350 watts per square metre (W/m^2): this is called the 'solar constant'. However, not all this energy reaches the Earth's surface. The atmosphere absorbs and reflects much of it, and by the time it reaches the Earth's surface, it is reduced to a maximum of about 1000W/m^2 (see Figures 2.2 and 2.3). This means that when the sun is directly overhead on a sunny day, solar radiation is

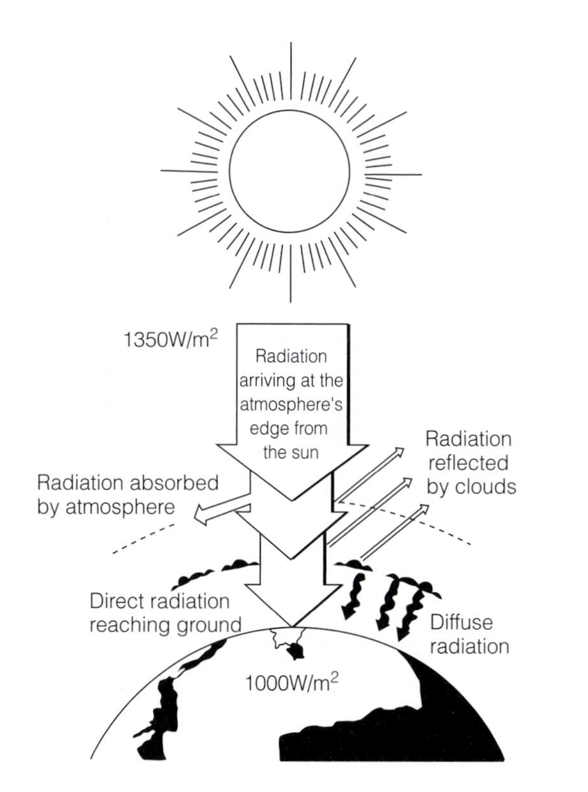

Figure 2.2 *Absorption and reflection of solar radiation by the Earth's atmosphere*

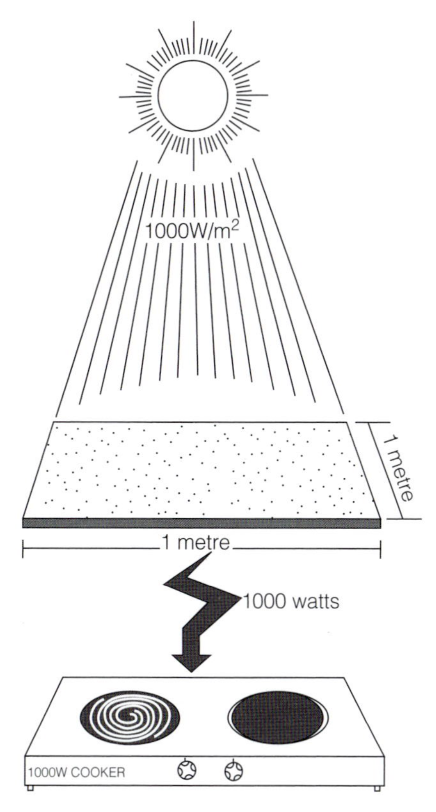

Figure 2.3 *How much power is 1000W/m²?*

arriving at the rate of about 1000W/m². Northern countries (i.e. Europe) have lower annual solar radiation levels than countries nearer the Equator – mainly because they have shorter days in winter.

Direct and Diffuse Radiation

Solar radiation can be divided into two types: direct and diffuse. Direct radiation comes in a straight beam and can be focused with a lens or mirror. Diffuse radiation is radiation reflected by the atmosphere or radiation scattered and reflected by clouds, smog or dust (Figure 2.4). Clouds and dust absorb and scatter radiation, reducing the amount that reaches the ground. On a sunny day, most radiation reaching the ground is direct, but on a cloudy day up to 100 per cent of the radiation is diffuse. Together, direct radiation and diffuse radiation are known as global radiation.

Radiation received on a surface in cloudy weather can be as little as one-tenth of that received in full sun. Therefore, solar systems must be designed to guarantee enough power in cloudy periods and months with lower solar radiation levels. At the same time, system users must economize energy-use when it is cloudy.

Annual and even monthly solar radiation is predictable. Factors that affect the amount of solar radiation an area receives include the area's latitude, cloudy

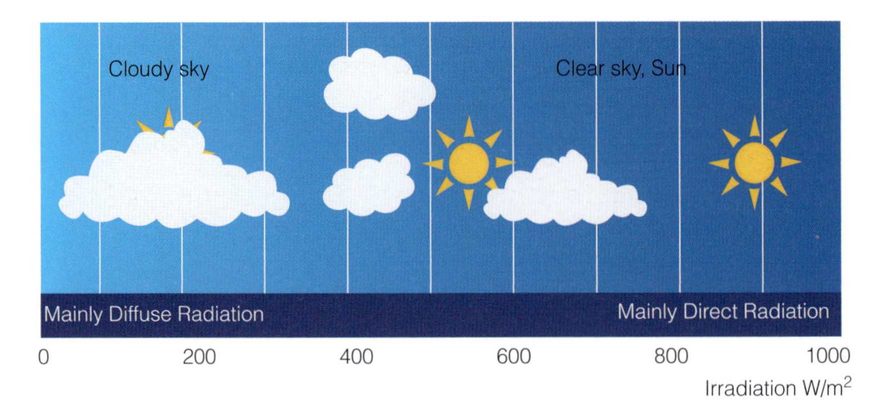

Figure 2.4 *Direct and diffuse radiation*

periods, humidity and atmospheric clarity. At high-intensity solar regions near the Equator, solar radiation is especially affected by cloudy periods. Long cloudy periods significantly reduce the amount of solar energy available. High humidity absorbs and hence reduces radiation. Atmospheric clarity, reduced by smoke, smog and dust, also affects incoming solar radiation. The total amount of solar energy that a location receives may vary from season to season, but is quite constant from year to year.

Solar Irradiance

Solar irradiance refers to the solar radiation actually striking a surface, or the power received per unit area from the sun. This is measured in watts per square metre (W/m^2) or kilowatts per square metre (kW/m^2). If a solar module is facing the sun directly (i.e. if the module is perpendicular to the sun's rays) irradiance will be much higher than if the module is at a large angle to the sun.

Figure 2.5 shows the changes in the amount of power received on a flat surface over the course of a clear day. In the morning and late afternoon, less

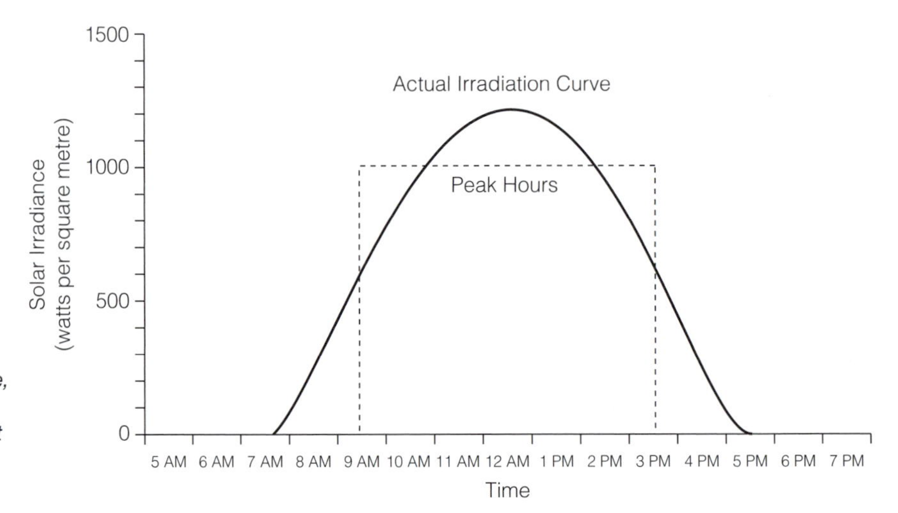

Figure 2.5 *Solar irradiance, in watts per square metre, received over time on a flat surface in an Equatorial region*

power is received because the flat surface is not at an optimum angle to the sun and because there is less energy in the solar beam. At noon, the amount of power received is highest. The actual amount of power received at a given time varies with passing clouds and the amount of dust in the atmosphere.

The angle at which the solar beam strikes the surface is called the solar incident angle. The closer the solar incident angle is to 90°, the more energy is received on the surface (see Figure 2.6). If a solar module is turned to face the sun throughout the day, its energy output increases. This practice is called tracking.

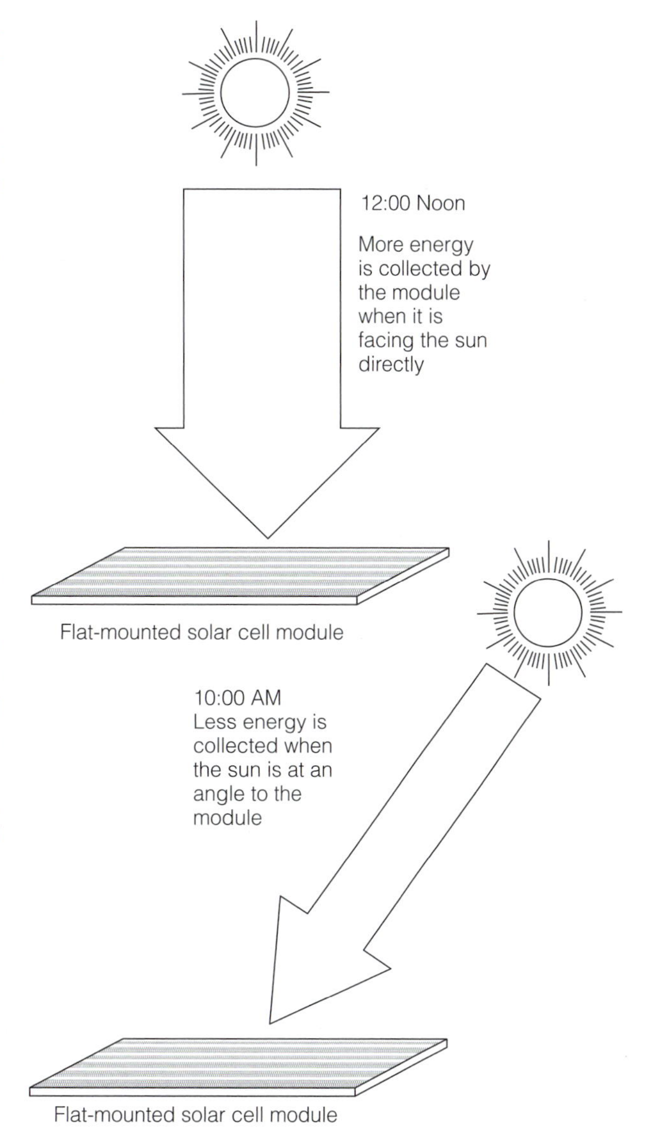

Figure 2.6 *The solar incident angle*

Insolation

Insolation (a short way of saying incident solar radiation) is a measure of the solar energy received on a specified area over a specified period of time. Meteorological stations throughout the world keep records of monthly solar insolation that are useful in planning solar utilization systems. For the purposes of off-grid solar system design, insolation is normally measured in either of two methods, as described in Table 2.1 below.

A site that receives 6 peak sun hours a day receives the same amount of energy that would have been received if the sun had shone for 6 hours at 1000W/m^2. In reality, irradiance changes throughout the day. At a good solar energy site, irradiance is above 1000W/m^2 for about 3 hours, between 800 and 1000W/m^2 for 2 hours, between 600 and 800W/m^2 for 2 hours and between 400 and 600W/m^2 for 2 hours and between 200 and 400W/m^2 for 2 hours. Still, the energy is equivalent to 6 hours of irradiance at 1000W/m^2 (Figure 2.5). For example, during October a site in Arusha, Tanzania, would be expected to receive 6.3kWh/m^2/day and 6.3 peak sun hours per day.

Table 2.1 *Insolation measurement*

Method	Abbreviation	Definition
Kilowatt-hours per square metre per day	kWh/m^2/day	Quantity of solar energy, in kilowatt-hours, falling on a square metre in a day.
Daily peak sun hours	PSH	Number of hours per day during which solar irradiance averages 1000W/m^2 at the site.

Peak sun hours are useful because they simplify calculations. They are commonly used when planning systems and are used throughout this book as the standard solar PV system planning measurement.

Figure 2.7 shows the mean daily insolation in peak sun hours for each month at four sites around the world. Note that the total amount of energy available per day changes considerably from month to month, even in Equatorial countries. On a sunny October day, Arusha, Tanzania, receives more than 6 peak sun hours of insolation. However, on a cloudy day in July the same site might receive only 4.3 peak sun hours.

Understanding and Using Meteorological Records

When planning a solar electric system, you will need to estimate your site's monthly mean daily insolation in kW/m^2 or peak sun hours. As a general rule, tropical locations receive between 3 and 8 peak sun hours per day. In the winter, northern climates receive less than 2 peak sun hours per day. The exact amount of insolation depends on the location and time of year. While it is difficult to accurately estimate how much solar energy a site will receive on any given day, it is possible to predict insolation fairly accurately on a monthly or annual basis.

Solar insolation is measured using a device called a pyranometer. For small systems, however, it is not necessary to buy or install an expensive pyranometer. Monthly daily insolation information is collected and kept by national and international agencies around the world. Insolation data for your site may be kept at a nearby meteorological station or at a government meteorological district office. It may also be available online (see Chapter 12).

Do not confuse 'insolation' data with 'mean daily sunshine hours' data (which weather stations collect). 'Mean daily sunshine hours' is a measure of the amount of hours that the sun shines (i.e. up to 12 hours a day), while 'peak sun hours' measures the actual amount of solar energy received (rarely more than 8kWh/m^2 per day). Mean daily sunshine hours may be useful as indicators

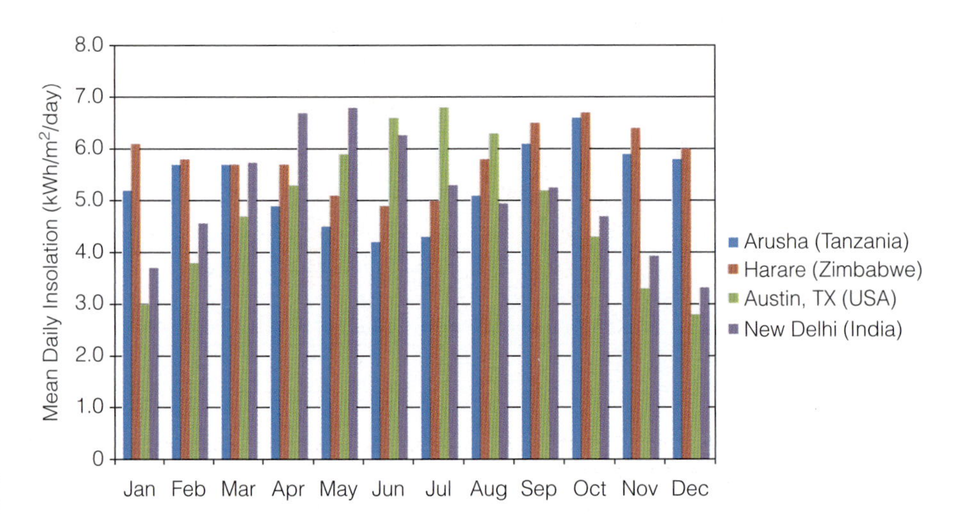

Figure 2.7 *Mean daily insolation in four cities*

if insolation records are not available for a site, but they should be carefully compared with national insolation records. For example, a site that receives '8 sunshine hours' might receive 4 or less peak sun hours.

The map in Figure 2.8 shows the lowest monthly daily insolation figures throughout the world. Based on long term meteorological records it gives a good indication of solar energy distribution in peak sun hours. Be careful using common capital city data for specific sites, as local weather conditions and site altitudes cause large variations in solar radiation in sites even a few hundred kilometres apart.

Angles, Solar Orientation and Tracking the Sun: Making the Best Use of Solar Energy

Solar radiation is absorbed most effectively when it strikes the solar collecting devices at a right angle (90°). For this reason, it is essential that solar energy collectors be correctly positioned to collect the maximum energy. The optimum mounting position varies with latitude and prevailing weather. When mounting solar collectors, keep in mind that, relative to the ground, the sun moves in a 180° arc from east to west each day. Also, its overhead north–south position varies greatly by season. For those south of the Equator, the sun is in the northern portion of the sky most of the time – solar panels should

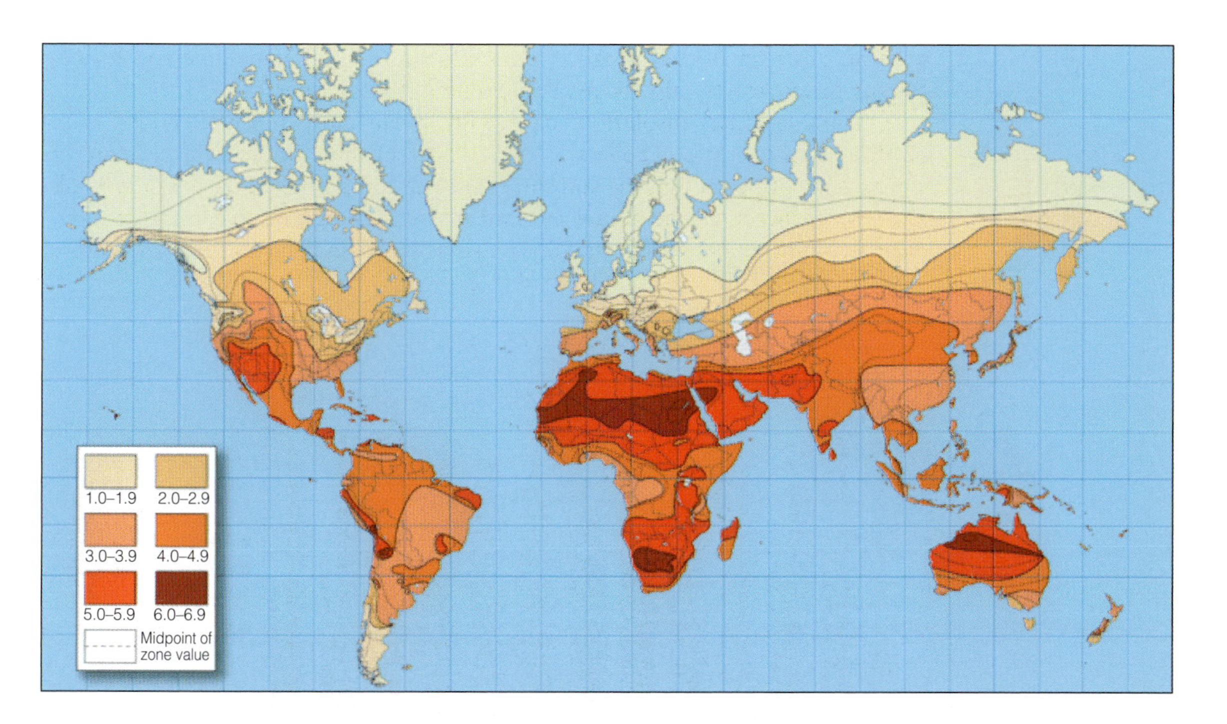

Figure 2.8 *Global insolation map. This shows the amount of solar energy (in hours) received each day on an optimally tilted surface during the worst month of the year.*

Source: www.altestore.com, based on accumulated worldwide solar insolation data

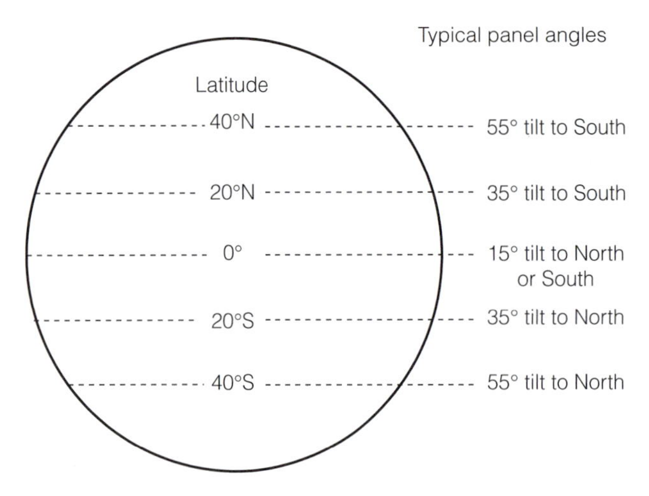

Figure 2.9 *Rule-of-thumb mounting angles for fixed solar collectors where the design month is the least sunny month, and the system relies 100 per cent on solar electricity generation*

be mounted with a tilt facing north. For those north of the Equator, the sun is usually in the southern portion of the sky and panels should be mounted facing south.

A rule of thumb is to mount modules facing the Equator at an angle equal to the site's latitude plus 10°. In Zimbabwe, for example, modules should be mounted facing north at 25–30°. Local solar dealers should be able to suggest the best tilt and direction for module-mounting in your area (Figure 2.9).

Along the Equator, the sun's position varies from 23° north to 23° south over the course of the year; modules should be mounted with a 5–10° north or south tilt towards the part of the sky (north or south) where the sun is during the least sunny months.

More solar power can be collected if the solar panels are turned to face the sun throughout the day. Turning solar devices to face the sun as it moves across the sky is called tracking. Tracking can increase the output of a solar panel by 30 per cent.

PV system installers may choose to use solar tracking devices that turn the solar array to face the sun automatically over the course of the day (see Chapter 12 for more information). However, for systems below 500 Wp (watt-peak) in size, the added expense of a solar tracker is rarely justified. Also, trackers add moving parts that can break down – not a good idea in very remote areas.

Sometimes, it is effective to reposition the solar module just a few times per day. A simple, cost-effective tracking mount can be made by attaching the array to a rotatable pole mount (making sure that someone turns the pole twice a day!) (see Figure 2.10).

Outside of the tropics, seasonal adjustments of the array angle can also gain considerable extra energy. During winter months, the array angle should be adjusted so that it is steeper; during summer months, this angle should be reduced.

Figure 2.10 *Simple manually turned pole-mounted solar tracker*

Basic Solar Energy Principles

As mentioned previously, we are able to harvest solar energy in two forms: as heat and as electricity. This section focuses mostly on principles of solar thermal technologies, which are often different from solar electric technologies. For solar thermal applications, radiant energy is directed using reflectors and concentrators. It is collected as heat by trapping and absorbing it on specially treated surfaces and by using the 'greenhouse effect'.

Principles of solar thermal technologies are provided herein for general information. It is good practice to understand the difference between solar thermal and solar electric technologies. Also, because solar electricity is not practical for generating heat, many off-grid PV users prefer to heat water and cook using solar thermal technology. Finally, some solar electric equipment also uses principles more commonly associated with solar thermal technologies.

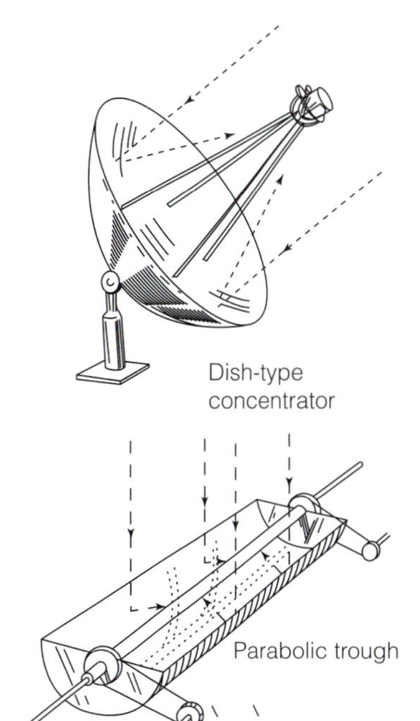

Figure 2.11 *Solar concentrators and reflectors*

Concentrating solar energy

Solar concentrators and reflectors (see Figure 2.11) help overcome the spread-out nature of solar energy. They redirect solar radiation incident over a large area, reflecting or focusing it to a small area where it is trapped or harnessed at high temperatures.

Concentrators make use of reflective surfaces and shapes that 'bounce' radiation off their surfaces and direct it to a focal point. By covering a large area with solar concentrators, most of the total radiation that falls on that area can be focused into a small area and absorbed as heat. Concentrators are useful only with direct (rather than diffuse) solar energy. They must be moved constantly to follow the changing position of the sun in the sky. Concentrated solar power (CSP) stations do this on an industrial scale, collecting and converting solar heat to electricity that is sold into the grid.

'Parabolic dishes' are bowl-shaped mirrors that focus radiant energy upon one point where the energy is concentrated and utilized (some solar cookers use these). 'Parabolic troughs' focus radiation along a pipe that runs through the centre.

'Simple reflectors' are flat mirrors or polished aluminium surfaces arranged around a device to increase the radiation incident upon it. Solar cookers and concentrating-type solar electric systems commonly use these.

Trapping solar energy

Solar energy is trapped as heat using properties of heat absorption, heat transfer, insulation and the 'greenhouse effect'.

Solar radiation is absorbed on surfaces as heat. The amount of heat radiation absorbed depends upon the incident surface area, the colour and material of the surface, and the angle and intensity of the incoming radiation. Solar energy is best absorbed on surfaces that are perpendicular to the incoming solar radiation.

Dark-coloured (i.e. black), non-reflective metal surfaces absorb solar radiation best. Solar water-heaters, stills, driers and cookers make use of black-painted absorber surfaces for capturing heat.

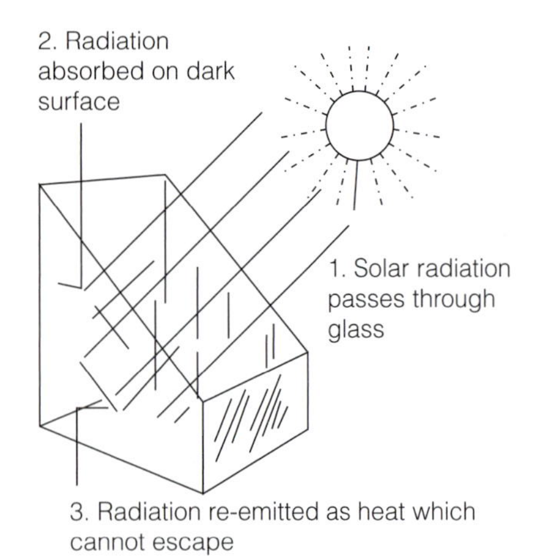

Figure 2.12 *The greenhouse effect*

The greenhouse effect (see Figure 2.12) uses the properties of light and heat energy to accumulate heat from solar energy. Radiant energy can pass through glass (or clear plastic) surfaces, but infrared (heat) energy cannot. Thus, solar radiation passes through the glass windows of a sealed box, is absorbed by the surface behind the glass and re-radiated as heat which cannot pass back through the glass. This heat is trapped inside the solar collector and causes the temperature to rise inside the box. Solar box-cookers, solar driers, solar stills and solar-heated homes use the greenhouse effect to gain heat. Flat plate collectors, used in solar water-heating systems, take advantage of both the greenhouse effect and absorption principles to trap solar energy for heating air and water.

Solar heat, once collected, is either stored or used immediately. With solar water-heaters, captured heat is stored in the water, which is transferred by pipe to insulated water tanks.

Solar Thermal Technologies

Solar crop driers

Solar driers heat air with collected solar radiation and use it to dry crops. There are two parts in the solar drying process. First, solar radiation is captured and used to heat air, increasing its ability to hold and carry water vapour. The second part of the process is the actual drying, during which heated air moves through the product, warming it and extracting moisture. Drying takes place in a large box called the 'drying chamber'. Air is either heated in a flat plate collector or through a window in the drying chamber.

- Traditional unimproved solar driers place drying products on racks that expose them to sunlight and wind. They dry crops at the normal outside temperature and humidity as solar radiation and the wind combine to dehydrate the crop.
- Direct solar driers are closed, insulated boxes inside which both solar collection and drying take place. Radiation is collected by the greenhouse effect through a transparent cover in the drying chamber that contains trays to hold the drying product. Heated air circulates through or above the product, removes moisture and carries it out through vents.

Table 2.2 *Examples of established uses of solar thermal energy*

Application	Method	Location
Salt production	Evaporation ponds	Worldwide
Solar water-heating	Flat plate collectors, Evacuated tubes	Worldwide
Solar crop and fruit drying	Indirect, direct driers	Worldwide
Concentrated solar power stations	Parabolic trough concentrators, concentrating towers	US, Europe

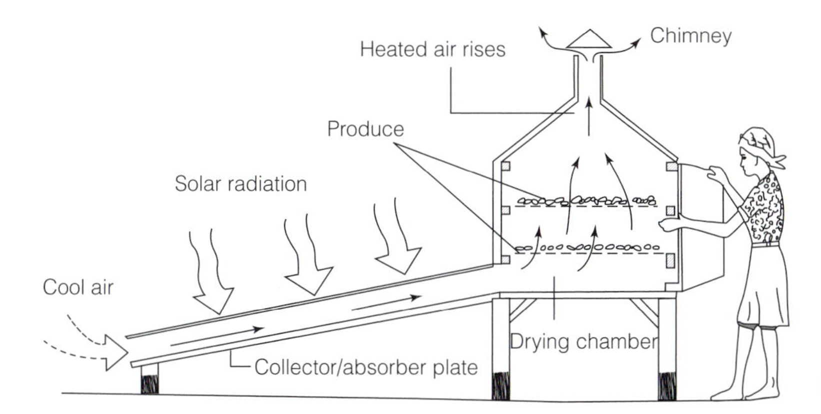

Figure 2.13 *Indirect solar drier*

- Indirect solar driers have a flat plate collector and a separate drying chamber that allow efficient use of solar energy and more control over the drying process. The solar collector heats air and, by convection, forces it through racks of drying products in the drying chamber (Figure 2.13).

Solar water-heaters

Solar water-heaters (SWHs) raise the temperature of water for bathing, washing, cleaning and cooking. The collectors consist of glass-covered panels (or evacuated tubes) with dark-coloured pipes inside. Water (or a heat transfer fluid) flows through the pipes and is warmed by the sun, then stored in insulated tanks for use in washing and bathing (Figure 2.14).

Figure 2.14 *Solar water-heater in Maasai Mara, Kenya*

Source: Mark Hankins

There are several different types of SWHs, but they can be roughly classified as follows:

- Batch-type: SWHs that heat and store the water in the same vessel but can only heat or keep a limited amount of water.
- Thermosyphon: SWHs that use a heating panel and an insulated storage unit which may also contain an electrical heating element for use during cloudy spells or periods of high hot-water demand. Thermosyphon SWHs do not have pumps but rely on the fact that hot water rises to move the solar-heated water from the solar collector to the storage tank – so the tank needs to be higher than the collector.
- Indirect: SWHs that are similar to thermosyphon-type SWHs but use a working fluid (usually glycol) to transfer heat from the panel to the storage unit. This overcomes problems of freezing water and corrosion in the panel. They usually have pumps.

Solar cookers

Solar cookers concentrate and collect solar energy as heat for preparing meals. They can be divided into two general groups (Figure 2.15 shows a box-type cooker that uses aluminium reflectors):

'Box-type cookers' are the most simple, least expensive and successfully disseminated solar cookers. A solar box-cooker is a large insulated box containing an enclosed cooking space with a window. One or more mirrors

Figure 2.15 *Solar cooker in Kenya, combining features of box and concentrator units*

Source: B. Wagner

(or aluminium sheets) reflect radiation into the box through the window. Food is placed inside the cooking space. When the window is closed and the cooker faces the sun, the inside cooking temperature may rise above 160°C (320°F). Box-cookers can be cheaply made and are best used for slow-cooking meals such as stews, porridge and rice dishes or for baking.

'Concentrating-type cookers' concentrate and focus solar energy as heat on to the vessel containing the food. Food is placed in the cooking vessel during operation, the vessel is placed on the cooker grill and the concentrator is adjusted to focus the sun's rays on to the pot. If the concentrator is properly oriented, the cooking vessel's bottom immediately heats up. Direct solar cookers produce temperatures high enough to roast or fry. Water reaches boiling point just as quickly as it does using conventional cookers. Some very large concentrating-type cookers are used to generate steam which then cooks the food in pressurized vessels.

The main advantage of solar cookers is that they make use of an abundant solar energy resource, but this must be weighed against several disadvantages:

- With most solar cookers, food can only be cooked during the hottest parts of the day and only on sunny days.
- Food takes longer to cook with most solar cookers.
- Most solar cookers require those who prepare meals to change or learn new cooking habits.
- Not all foods can be cooked with solar cookers.

Box 2.2 Energy efficiency

A basic understanding of energy efficiency is important when planning solar energy systems. 'Efficiency' is the ratio of output energy to input energy. In the case of solar energy systems, the input energy is the energy received from the sun by the modules or collectors. The output energy is the electricity available for lights and appliances (with solar electricity) and heating (with solar thermal). Efficiency is measured as a percentage; the higher the percentage, the more efficient the energy transfer.

A major problem of solar electric energy conversion and utilization is the low conversion efficiency of PV equipment. Although the sun is shining at a rate of 1000W/m^2, with most solar equipment we cannot use all of that power because of efficiency losses. Solar thermal devices generally have a higher efficiency than solar electric devices, but their output is heat, not electricity. As a form of energy, heat is of lower quality than electricity – it is harder to store and cannot be used for as many applications (for example, you cannot power a TV with heat energy – you need electricity!). This is the reason solar electricity is so popular despite its higher cost.

To give an example of energy efficiency: at noon on a sunny day, Equatorial solar radiation arrives at a rate of about 1000 watts per square metre (this is 'power') (see Figure 2.3). If all of the solar radiation striking 1 square metre in 1 day in an Equatorial location could be collected, then a total of 6 kilowatt-hours could be harvested per square metre of solar cell module per day (this is 'energy'). However, it is impossible to capture all of the solar energy because of efficiency losses (i.e. energy is lost as heat or in reflections). The best commercial solar cell modules have an efficiency of less than 18 per cent, which means that they can only transform about one-fifth of the arriving radiation into electricity (at noon, they transform only 200W/m^2 of the incoming 1000W/m^2 into electric power). Over the course of a day within the tropics, the energy converted to electricity by 1 square metre of very efficient solar cells is about 1000 watt-hours, or enough to power five 20W lamps for 10 hours.

Because of the high cost of the energy, solar electric systems should always use energy-efficient appliances. Table 2.3 gives the approximate efficiencies of several solar energy technologies.

Table 2.3 *Solar energy technologies and approximate efficiencies*

Solar application	Type of technology	Output energy	Approximate energy conversion efficiency
Solar water-heater	Flat plate collector	Heat	50–85%
Solar cooker	Concentrator type	Heat	70–85%
	Box type	Heat	30–50%
Solar drier	Flat plate collector	Heat	25–50%
Amorphous PV Cell	Amorphous silicon	Electricity	3–8%
Crystalline PV Cell	Crystalline silicon	Electricity	10–20%

3
Solar Cell Modules

This chapter gives general details about solar cell modules, also called photovoltaic or PV modules. It describes basic principles by which solar cells operate, the types of solar cells and the modules available. The energy output and characteristics of modules under various temperature, radiation and weather conditions are explained. Module ratings and the I-V curves (current-voltage curves) are described. The information in this chapter will help system designers to choose appropriate modules and estimate module energy output according to their local conditions.

Solar Cells and the Photo-electric Effect

The photo-electric effect

The photovoltaic is the direct conversion of sunlight to electricity. Light striking solar cells is converted into electric energy. This occurs according to a principle called the 'photo-electric effect'. Solar electric devices are also called photovoltaic or PV devices.

The photo-electric effect was first discovered during the 1830s by a French physicist, Alexandre Edmond Becquerel, who noted that certain materials produce electric current when exposed to light. The amount of power produced, however, was insignificant and the property did not find a useful power-generating application until the 1950s, when the United States space programme decided that solar cells would be a good electrical power source for satellites. Following the oil crisis in 1973, much research was put into improving PV technology for applications on Earth. The first automated production of solar cells began in the US in 1983. By 1990, the industry was well established and growing. Today solar cells are manufactured in many countries, led by China, Germany, Japan, Spain, the US and India. Figure 1.1 shows the increasing trend for world shipments of PV modules since the mid-1990s.

Solar Cells and Solar Cell Technology

The basic unit of solar electric production is the solar cell. Light striking solar cells creates a current powered by incoming light energy (see Figure 3.1). They produce electricity when placed in sunlight. Most solar cells do not get used up or damaged while generating electric power. Their life is limited only by breakage or long-term exposure to the elements. If a high-quality solar cell module is properly protected, it should last for more than 25 years.

Box 3.1 How solar cells work: a basic explanation

Solar cells rely on the special electric properties of the element silicon (and other semiconductor materials) that enable it to act as both an insulator and a conductor. Specially treated wafers of silicon 'sort' or 'push' electrons dislodged by solar energy across an electric field on the cell to produce an electric current. Other materials are also used but the vast majority of solar cells are made from silicon.

Solar radiation is made up of high-energy sub-atomic particles called photons. Each photon carries a quantity of energy (according to its wavelength); some photons have more energy than others. When a photon of sufficient energy strikes a silicon atom in a solar cell, it 'knocks' the outermost silicon electron out of its orbit around the nucleus, freeing it to move across the cell's electric field, also called the p-n (positive-negative) junction. Once the electrons cross the field, they cannot move back. As many electrons cross the cell's field, the back of the cell develops a negative charge.

If a load is connected between the negative and positive sides of the cell (see Figure 3.1), the electrons flow as a current. Thus, solar energy (in the form of photons) continuously dislodges silicon electrons from their orbitals and creates a voltage that 'pushes' electrons through wires as electric current. More intense sunlight gives a stronger current. If the light stops striking the cell, the current stops flowing immediately. (For more detailed information about the photo-electric effect, consult the resources in Chapter 12.)

A number of commercial varieties of silicon-type solar cells and solar cell modules are available. When choosing modules, it is important to consider the advantages and disadvantages of the various types. Monocrystalline and polycrystalline silicon together accounted for about 90 per cent of solar PV production in 2008. Monocrystalline refers to cells cut from single crystals of silicon (a crystal is the regular geometric state taken up by elements in certain conditions; similar silicon crystals are specially grown for the computer industry). Polycrystalline refers to cells made from many crystals. Amorphous silicon cells are made from non-crystalline silicon that is deposited on the back of glass or other substrates. Thin film PV technologies (which include amorphous silicon) include a number of other technologies as explained below. The PV industry is growing rapidly and new types of technologies become available each year. Table 3.1 describes PV technologies that you are likely to encounter in the market.

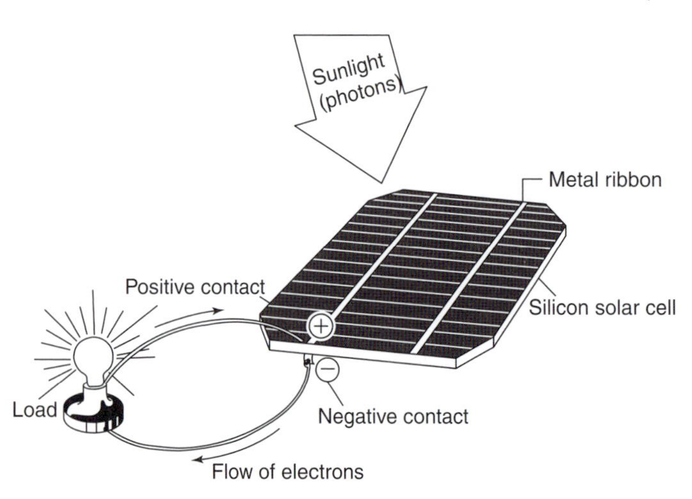

Figure 3.1 *The photo-electric effect*

Modules are less efficient than cells mainly because there are gaps between and around the cells of a module. Data given here refers to laboratory efficiencies that are always higher than efficiencies achieved by commercially available PV modules.

Table 3.1 *Commercially available solar cell module types*

Type of PV technology	Maximum cell efficiency	Typical commercial module efficiency	Notes
Crystalline Silicon			
Monocrystalline	24%	11–17%	Fully mature technology: 35% of world production (2007)
Polycrystalline	20%	11–15%	Fully mature technology: 45% of world production (2007)
Ribbon	19%	7–13%	Fully mature technology
Thin Film			
Amorphous Silicon	13%	4–8%	Initial degradation in performance
Multi-junction Amorphous Silicon	12%	6–9%	Similar to Amorphous Silicon Flexible
Cadmium Telluride	17%	7–8.5%	
Copper Indium Gallium Di-Selenide (CIGS)	19%	9–11%	
Organic (Dye)-type solar modules	12%	3-5%	Relatively uncommon
Other Types			
Hybrid HIT	21%	17%	Combined Amorphous Silicon and Crystalline

Source: Planning & Installing PV Systems, various others

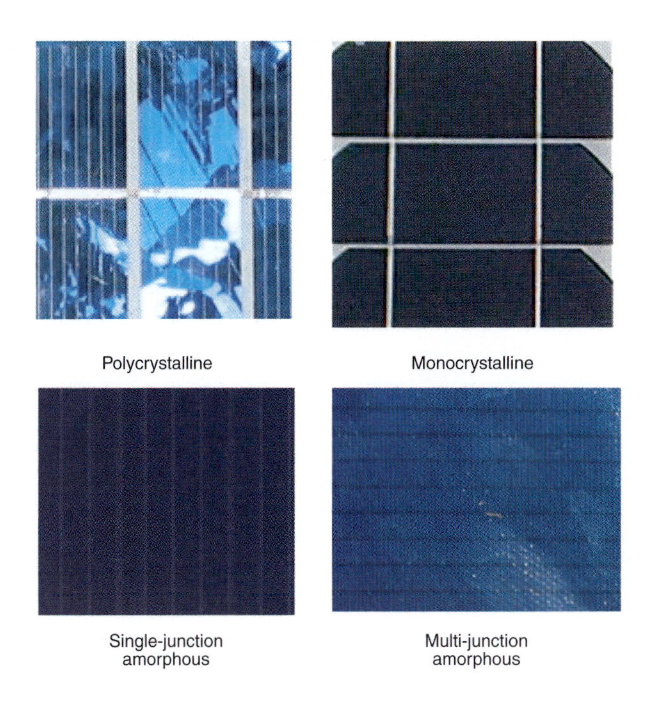

Figure 3.2 *Various types of silicon solar cells*

Crystalline PV technologies

Monocrystalline and polycrystalline solar cells are specially treated wafers sliced from cylindrical silicon crystals using very precise saws (in the process of slicing much of the expensive crystal is lost as dust). These thin wafers are chemically treated (or 'doped') in furnaces to enhance their electric properties. Anti-reflective coatings are applied to cell surfaces to help them absorb radiation more effectively. After this, thin metal wires are soldered to the front of the cell. These 'ribbons' of metal on the cell act as the positive contact, whereas a solid layer of metal on the back of the cell acts as a negative contact (see Figure 3.1).

Monocrystalline cells were the first to be developed for commercial purposes. Production cells have a laboratory efficiency of up to 21.5 per cent (i.e. if solar radiation is striking the cells at a perpendicular angle with an intensity of $1000W/m^2$, about $215W/m^2$ of solar energy is converted to electricity). Monocrystalline cells are chemically stable, so they last for a very long time if properly protected.

Polycrystalline (or multicrystalline) cells have a slightly lower laboratory efficiency than monocrystalline cells (up to 20 per cent). Like monocrystalline cells, they have a long lifetime and do not degrade. They are sliced from cast ingots of polycrystalline silicon (made by a different process than monocrystalline silicon). Whereas monocrystalline cells have a single colour tone, polycrystalline cell surfaces have multiple patterns. As of 2008, the price per watt of both types of crystalline cells was about the same per peak watt output.

Thin film modules

Thin film modules use non-crystalline PV material that can be deposited in fine layers on various types of surfaces. Although they only accounted for about 10 per cent of all solar PV production in 2008, their portion of the market is growing rapidly. It is expected that thin film technology costs will drop much faster than that of crystalline silicon technology.

Amorphous silicon cells are the most common thin film PV technology. Amorphous-type cells do not use silicon in crystalline form; instead, high temperature silane (SiO_4) gas is deposited as very thin layers of amorphous silicon on the back of a glass or plastic substrate. The surface is then scored to divide it into cells and electrical connections are added. Amorphous silicon cells are commonly used in toys, calculators, garden lights and watches; 10–20Wp amorphous modules are common in developing country markets.

Amorphous cells operate according to the same principles as crystalline modules, but have much lower efficiencies (4–8 per cent). Because they operate at low efficiencies, amorphous modules must be three to four times the size of monocrystalline or polycrystalline modules to generate the same power.

Amorphous silicon degrades over the first few months of exposure to the sun. Thus, when first installed, new modules produce as much as 25 per cent more power than their rating; after several months they degrade down to their rated power output.

Multi-junction cells utilize two or more layers of amorphous films deposited on one surface to collect a higher portion of the solar radiation. Their efficiency may be twice as high as single junction amorphous cells. Some multi-junction cells (and modules) have another advantage of being flexible and less breakable (a common process deposits the PV material on flexible metal sheets). This is an important point to keep in mind if the modules are being transported over rough roads in remote areas.

Cadmium telluride (CdTe) and copper indium gallium di-selenide (CIGS) cells are two types of thin film technologies that are increasingly available. CdTe cells are more efficient than amorphous silicon and can be produced at a lower cost. CIGS cells are the most efficient of all thin film technologies and, like CdTe cells, are not subject to degradation.

Other types of solar cells

Concentrator cells are smaller area cells that convert solar energy focused on them by reflectors or Fresnel lenses. Designed to efficiently produce power at high temperatures, these monocrystalline cells (some other solar cell materials are also used) use less solar cell material per output, so they are less expensive to produce. Unlike other PV modules, however, they cannot use diffuse solar energy. To date, only a small portion of the solar market is made of concentrator cells, but many types are now commercially available.

Hybrid cells (Heterojunction with Intrinsic Thin layer or HIT) combine a crystalline cell with thin film cells to increase electric output. They have outputs as much as 7 per cent higher than the best polycrystalline modules and they have better performance at high temperatures.

Dye-sensitized cells (or Grätzel cells) are based on the semiconductor titanium dioxide. They use an organic dye to capture solar energy (in a process somewhat similar to the way chlorophyll captures light in plants). Although dye-sensitized cells are still in the early stages of development and production, they use non-toxic materials, they are cheap to produce and they perform well in high temperatures and low light intensities.

Solar Cell Modules and Arrays

As mentioned in the previous sections, solar cells are made from a variety of materials. They also vary greatly in size and are carefully selected to meet the electrical needs of the load. The amount of current produced by a solar cell depends on its size and type. A $10cm^2$ ($1.55\ inch^2$) monocrystalline cell produces a current of about 3.5 amps under Standard Test Conditions (see Box 3.2). A similarly sized amorphous silicon cell would produce about one-third this amount.

No matter the task, though, all silicon-type solar cells generate a potential difference (voltage) of between 0.4 and 0.5 volts in normal operation. For this reason, solar cells must be connected in series to increase voltage to a useful level. For most tasks, it is not convenient to use single solar cells because their output does not match the load demand. For example, one cell cannot power a radio if the radio requires current at 3 volts and the cell produces a voltage of only 0.5 volts. Five cells in series are enough to power a calculator of 2 volts and 36 cells are normally required to charge a 12 volt battery.

Box 3.2 Standard Test Conditions

Standard Test Conditions (STC) enable manufacturers and consumers to compare how different PV devices perform under the same conditions. A number of laboratories and centres test modules and cells for manufacturers.

Standard test conditions are set at:

- 1000W/m^2 solar irradiance;
- 25°C;
- air mass of 1.5 (air mass indicates how much radiation is absorbed by the atmosphere).

Note that modules and cells almost always produce *less* power under actual working conditions.

Thus solar cells are arranged in series to increase voltage and the number of cells depends on the application. Moreover, crystalline solar cell wafers are fragile, so they must be protected from breakage and corrosion. For these reasons, solar cells are electrically connected in series, then packaged and framed in devices called photovoltaic modules.

From solar cells to photovoltaic modules

Arrangements of many solar cells wired in series, sealed between glass and plastic, and supported inside a metal frame are called 'photovoltaic modules'. Groups of modules mounted together are called 'arrays' (Figure 3.3).

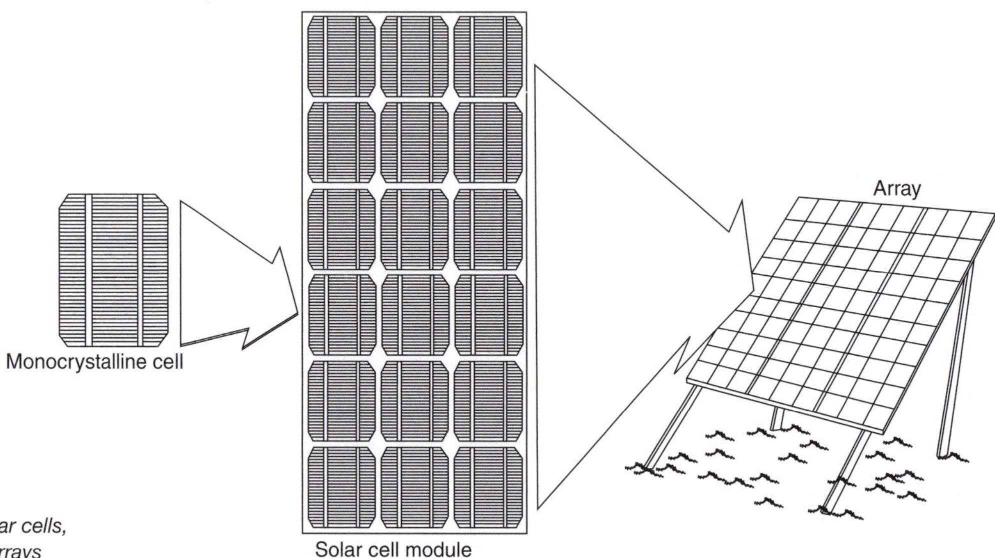

Figure 3.3 *Solar cells, modules and arrays*

The process of making solar cell modules from monocrystalline and polycrystalline silicon cells involves several steps. Once properly prepared and treated with anti-reflection coatings, solar cells are soldered together in series (i.e. the front of one cell is connected to the back of the next) and then mounted between glass and plastic. The process by which monocrystalline or polycrystalline solar cells are sealed between glass and plastic is called 'encapsulation'.

During encapsulation, cells are sealed at high temperature between layers of plastic (a special type called EVA plastic) and glass in such a manner that air or water cannot enter and corrode the electrical connections between the cells. Modules are then cased in metal or plastic frames to protect their edges and to protect them from twisting. The frame may have holes drilled in it for easy mounting and a connection terminal for earthing/grounding cables.

Positive and negative electric contacts from the cells, either terminal screws or wires, are fixed on to the back of the module. With most large modules (i.e. above 40Wp), the terminals are enclosed in a junction box. Some modules have junction boxes arranged so that their voltage can be tapped at several different voltages (e.g. 6 or 12V).

With thin film modules, 'cells' are not individual wafers, but individual parts of the substrate material that have been separated from each other by scored divisions. See Figure 3.7 to examine the difference between a crystalline module and an amorphous silicon module and look at how the 'cells' are different from each other. Because many types of modules look similar to each other, you should always look at datasheets and labels to be sure which type of module you are getting.

Module ratings

All solar cell modules are rated according to their maximum output, or 'peak power'. Peak power, abbreviated Wp (watt-peak), is defined as the amount of power a solar cell module is expected to deliver under Standard Test Conditions (STC). The module's power rating in peak watts should be specified prominently on the module's label by the manufacturer or dealer. Modules almost always produce less power than their rated peak power in field conditions.

Solar cell modules are available in various voltages. Some modules for grid-tied purposes are rated as high as 72 volts, whereas modules for special lighting and communication tasks may be 3 volts or less.

Most off-grid PV systems – the type covered in this book – use modules with 36 cells (sometimes 72 cells). Thirty-six-cell modules produce the best voltage for charging a 12V battery. When buying, always check the module's rated power and rated voltage!

By counting the number of cells, it is possible to estimate a module's voltage. Remember, for charging 12V batteries in most of the types of stand-alone systems discussed in this book, crystalline silicon modules should have 36 solar cells wired in series. Count the number of cells in your module before buying!

Arrays

Often, a number of modules are required to meet the power requirements at a site. When mounted together, groups of modules are referred to as arrays (Figure 3.3). For example, a solar health centre with a vaccine fridge might require an array of four to six 50Wp modules (i.e. 200–300Wp). A community solar water pump might require an array of fifteen to twenty 50Wp modules (i.e. 750–1000Wp).

Output of Solar Cell Modules

The power output of a module depends on the number of cells in the module, the type of cells and the total surface area of the cells.

The output of a module changes depending on:

- the amount of solar radiation;
- the angle of the module with respect to the sun;
- the temperature of the module; and
- the voltage at which the load (or battery) is drawing power from the module.

Planners of solar PV systems should keep these important operating principles of solar modules in mind. The first three points – the amount of radiation, the sun-module angle and the temperature of the module – should determine where and how the module is mounted. These points will be revisited in Chapters 8 and 9.

The last point relates to the voltage of the battery, the charge regulator and/or the load. Properly selected and configured modules (or arrays), charge regulators and loads (especially batteries) get the most power from the sun.

The I-V curve

I-V curves are used to compare solar cell modules and to predict their performance at various temperatures, voltage loads and levels of insolation.

Each solar cell and module has its own particular set of operating characteristics. At a given voltage, a module (or cell) will produce a certain current. These properties are described by the current-voltage curve, better known as the I-V curve.

Figure 3.4 shows an I-V curve for a module rated at 42Wp at STC. The left-hand side (I) gives the current the module produces depending on voltage. The bottom side gives the voltage produced by the module at various currents. At each point along the line, you can determine the power of the module by multiplying the current by the voltage. For example, imagine a battery being charged by a module: at 12 volts (Point A, Figure 3.4), the current from the module is 3.2 amps and the power output is 38.4 watts (amps × volts = power, so 12V × 3.2A = 38.4W, which is less than the rated 42Wp).

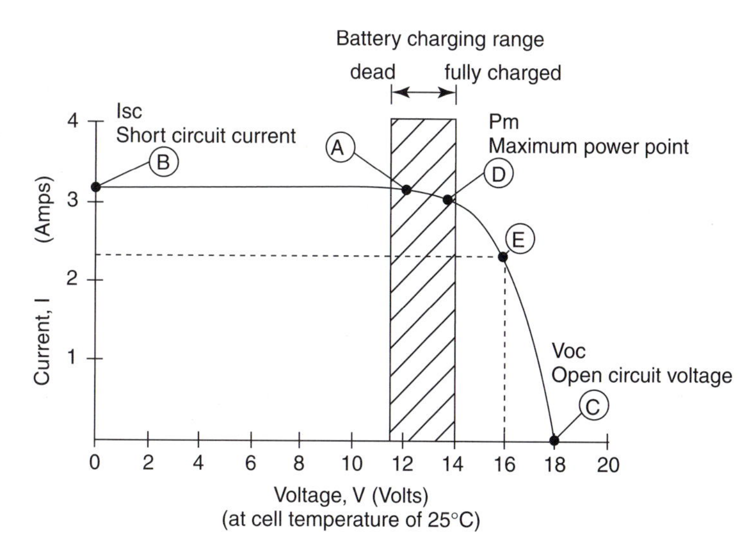

Figure 3.4 *The I-V curve*

There are several points of interest about the I-V curve in Figure 3.4:

- Isc, the short-circuit current, is the current measured in full sunlight when the positive and negative wires are 'shorted'. In practice, a suitably rated ammeter is attached to the positive and negative leads of the module. On the I-V curve, this is the point where the curve crosses 0 volts (Point B). This is the maximum current that the module is capable of producing.
- Voc, the open circuit voltage, is the voltage measured with an open circuit. It is measured with the module in full sunlight using a voltmeter attached to the positive and negative leads of the module. On the I-V curve, this is the point where the curve crosses 0 amps (Point C). This is the maximum voltage that the module can produce on a sunny day.
- Pm, the maximum power point, is the point on the I-V curve where the module produces the greatest power (its rated maximum, which in this example is 42Wp). The maximum power point is always found at the place where the curve begins to bend steeply downward ('the knee', Point D). It is advisable to operate a module as near to the maximum power point as possible. If, for example, the module in Figure 3.4 is operating a load that demands 16 volts (Point E), power output (at 36 watts) is much less than that at the maximum power at Point D.

In general, the closer the 'knee' of the I-V curve is to the shape of a square, the better output characteristics of the module. Crystalline modules have I-V curves that are more 'square' than thin film modules. Compare the I-V curves of the various modules in Figure 3.7.

The shaded portion of Figure 3.4 shows the voltage boundaries within which a lead-acid battery is charged. A battery in a low state of charge is close to 11V, a battery in a high state of charge may be above 14V. Note from the I-V curve that as a battery's charge increases, the charging current from the module begins to decrease.

Box 3.3 Effects of radiation intensity on module output

Solar cell module output is very much governed by the intensity of the solar radiation on a module. Figure 3.5 shows that module output is directly proportional to the solar irradiance. Halving the intensity of solar radiation reduces the module output by half. Lower radiation also lowers the voltage at which current is produced. Look at the I-V curves in Figure 3.5: a 50 per cent drop in insolation causes a 50 per cent drop in current.

Cloud cover reduces the power output of a module to a third or less of its sunny weather output. During cloudy weather, the voltage of a module is also reduced. In hot, cloudy weather modules charging 12V batteries should be selected so that they maintain a high voltage – make sure they have 36 cells.

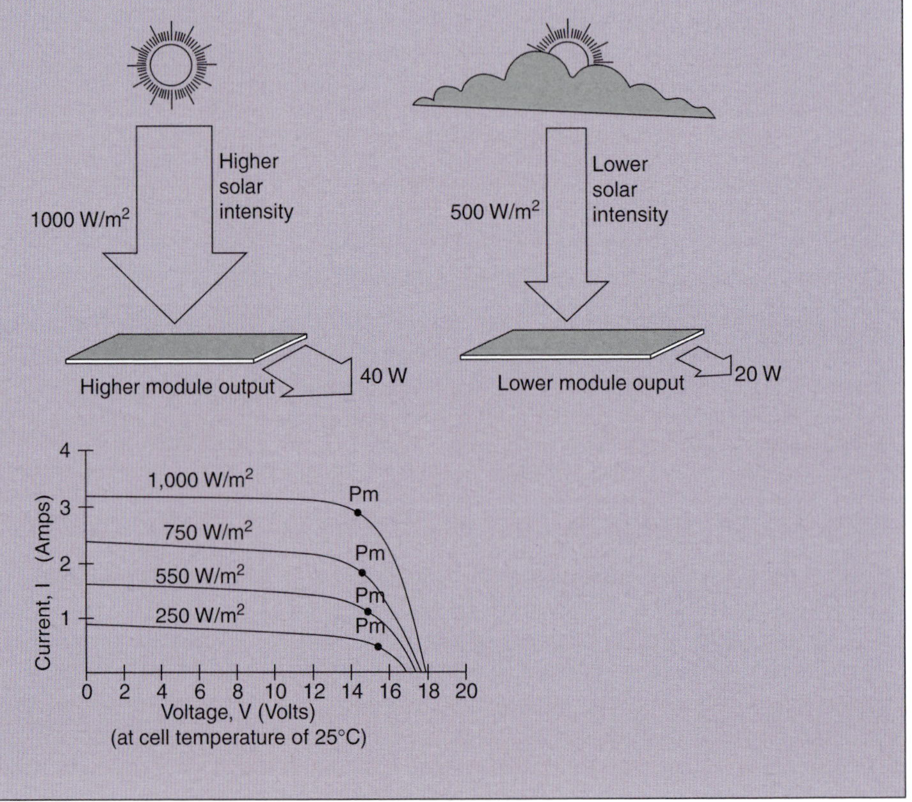

Figure 3.5 *Effects of radiation intensity on module output*

Effects of shading on module output

Obviously, if a shadow falls across all or part of a module, its electric output will be reduced. In fact, even shading a single cell can considerably lower a crystalline module's output and possibly damage it. Damage occurs because the cells in a module are connected in series and they each must carry the same current. When one cell (or more) is shaded, it stops producing current and instead consumes current, converting it to heat.

If a single cell is shaded for a long time, it may cause the entire module to fail. Even a single tree branch, a weed or a bird's nest can shade one cell and

Box 3.4 Effects of heat on module output

Unlike solar thermal devices, most solar PV modules produce less power as they get hotter. As the temperature increases, power output of monocrystalline solar cells falls by 0.5 per cent per degree centigrade (this is shown by the I-V curve in Figure 3.6). Thus, a 5°C (9°F) rise in temperature will cause a 2.5 per cent drop in power output. When mounted in the sun, solar cell modules are usually 20°C (36°F) warmer than the thermometer temperature. Note the differences in the I-V curves at various temperatures. At 60°C (140°F) the current 12V output of the module is much lower than at 10°C (50°F)!

This is important because the temperature on some rooftops can be higher than 60°C (140°F), reducing the output of the module by 20 per cent or more below its rated output. For this reason, installers are encouraged to mount modules on poles, on structures above the roof or in places where they are cooled by airflow to keep output as high as possible.

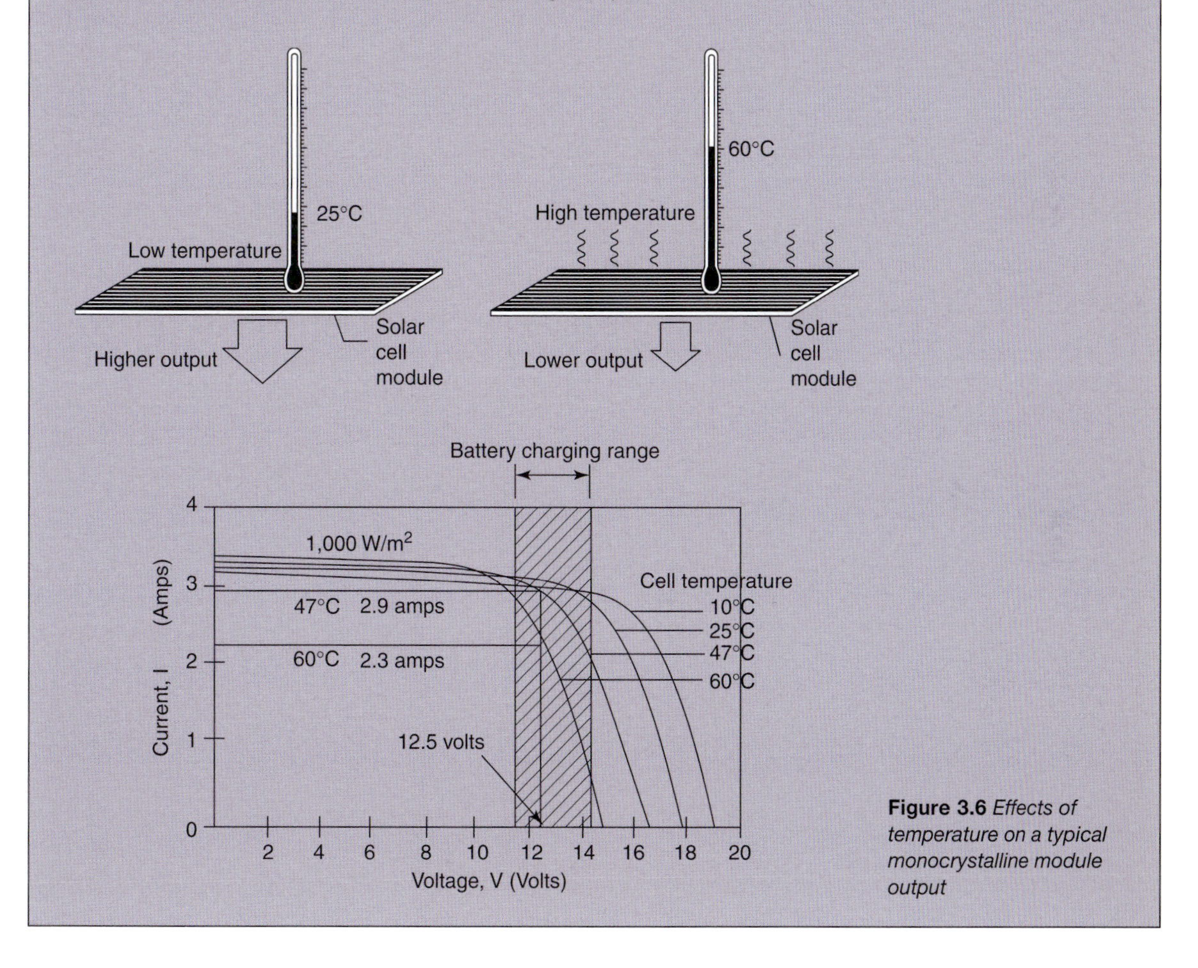

Figure 3.6 *Effects of temperature on a typical monocrystalline module output*

cause electrical production to fall dramatically. Amorphous and multijunction-type modules are less affected by small shadows than crystalline-type modules. Try covering one cell of a crystalline module when measuring the current output – you can easily see the difference!

Calculating a Solar Module's Output

The output of a module in field conditions can be estimated by multiplying the current produced by the module at load with the number of peak sun hours at the site. The operating current at load is used because, as mentioned in the I-V curve section, a module only produces its rated power at a particular voltage under standard conditions (it is not accurate to multiply the peak watt rating of a module by the peak sun hours).

Remember, unless they are installed in a cold, windy place, module operating temperature is far more likely to be between 40°C and 60°C (around 100–140°F) than the STC of 25°C (77°F). Also, when charging, modules do not always produce power at the maximum power point – in fact, they usually produce power 10–20 per cent less than their STC-rated power.

Suppliers usually provide 'typical' current output at 12–15V load for their modules. This is often called Normal Operating Cell Temperature (NOCT). Many suppliers provide I-V curves of their module's operation at NOCT; this is useful because it allows you to make a more accurate prediction of module output than at STC.

Choosing Solar Cell Modules

Modules should be chosen according to the energy requirements of the system load. Information on calculating the system load and planning systems is given in Chapter 9. Some important considerations when choosing modules include the following:

- If the module is crystalline, check the number of cells. Try to use modules with 36 cells, if possible. Never use crystalline silicon modules with fewer than 34 cells for charging batteries. However, using modules with more than 36 cells means, in effect, that the extra cells are not used in operation and may be a waste of money. Sometimes, 72-cell modules are used with a 24V battery bank.
- Make sure the module has a label that clearly states its peak power, the short-circuit current, the open circuit voltage, the current at load and where

Box 3.5 Example for calculating module output

Use the calculation below to predict your module's output on a typical day:

Operating current at load (amps) × peak sun hours (PSH) = Expected output of module (amp-hours/Ah)

When charging a battery at 12.6V, the current output of a module is about 3.1A. If the site receives 5 peak sun hours, then the module output is:

3.1A × 5psh = 15.5Ah

This figure can be multiplied by 12V to get watt-hours (Wh):

15.5Ah × 12V = 186Wh

the module was made. Also make sure that the module has an acceptable international certification such as IEC 61215 (for crystalline modules) or IEC 61646 (for thin film modules) (see Chapter 12 for more information on solar PV standards).

- Always check what type of guarantee the dealer offers. Modules should come with at least a 5-year warranty. Many good crystalline modules have 25-year warranties!
- If you are buying a second-hand module, it is a good idea to test the module using a multimeter – in full sunlight – before buying it. Compare its output with its rated output (if the dealer has the information). If you cannot test it yourself, find someone who can help you.
- There are many types of modules made for a number of applications! Make sure your module is suitable for charging 12V batteries and is suitable for stand-alone applications. As a general rule, modules made for grid-tied applications or specialized low-voltage applications should not be used (though there are exceptions).
- Shop around when looking for modules. Carefully compare features, prices and warranties. Always get copies of the I-V curves and datasheets of the modules you are thinking of buying and make sure the modules are suitable. Buying the wrong modules is an expensive mistake.
- Be careful about general and unspecific statements made by manufacturers or dealers about modules. For example, some amorphous module dealers state that their modules are 'better for tropical countries' because they are 'less affected by heat'. This is an over-simplification and can be misleading. When in doubt check the I-V curves of the module or ask an expert!

Box 3.6 Types and sizes of modules

Table 3.2 and Figure 3.7 give details on a number of types and sizes of modules commonly available. Note the different parameters and I-V curves.

Table 3.2 *Features of selected solar modules*

Feature	Sharp	SolarWorld	Free Energy Europe	Shurjo Energy
Model	Off-Grid	85-P (previously Shell)	FEE 20-12	SE85-A1S
Type	Polycrystalline	Monocrystalline	Amorphous Silicon	Thin film crystalline CIGS
Rated Output	80 Wp	85 Wp	19 Wp 16 Wp stabilized	85Wp
Current at Peak Power	4.63A	4.95A	0.99A	4.8A
Isc	5.15A	5.45A	1.22A	6.5A
Voc	21.6V	22.2V	22.8V	26.1V
Expected max output at NOCT (45.5°C, 800W/m^2)	58.8 Wp	63 Wp	N/A	59.8Wp
Number of Cells	36	36	30 scored sections	N/A

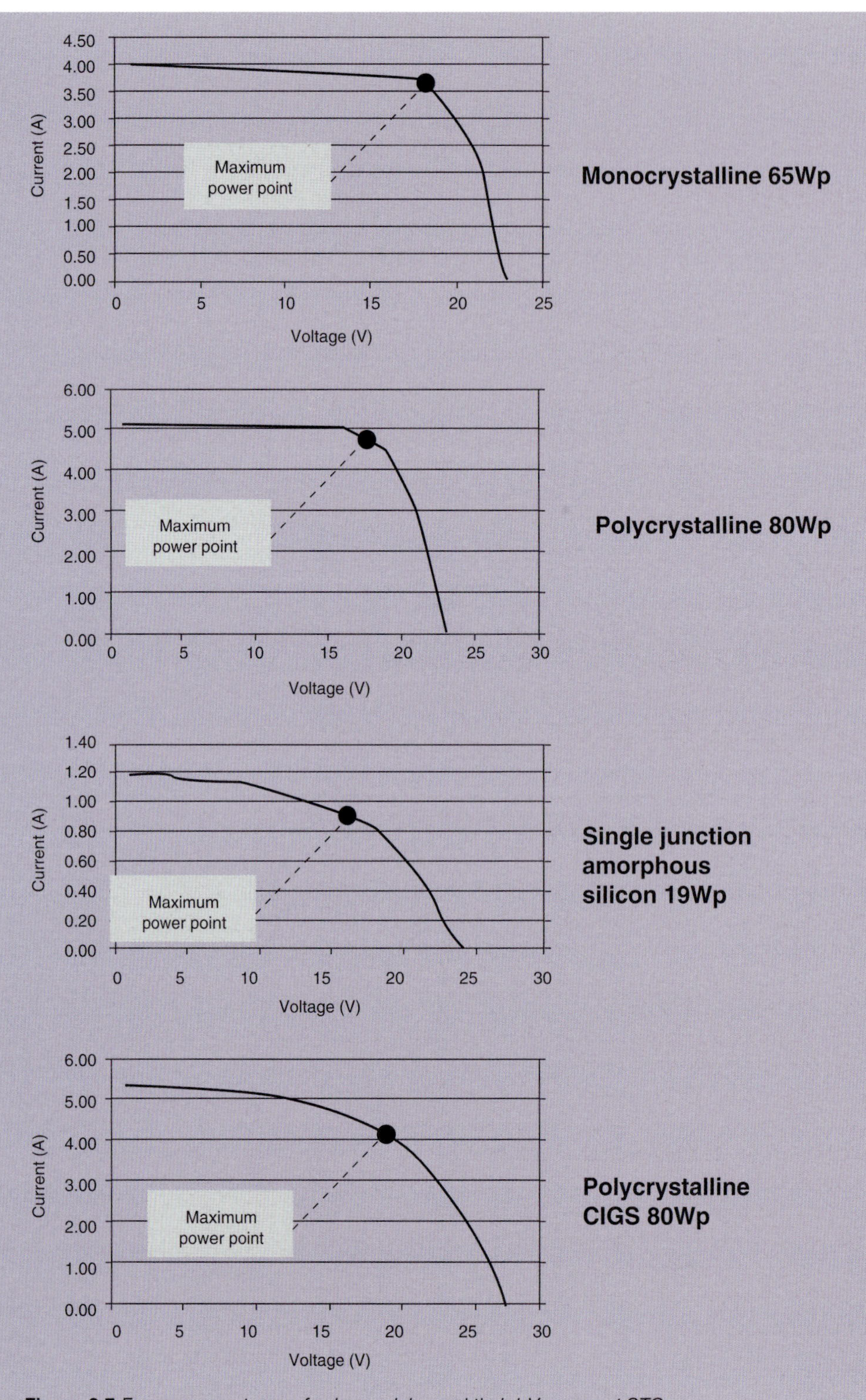

Figure 3.7 *Four common types of solar modules and their I-V curves at STC*

4
Batteries

This chapter provides information about rechargeable batteries that are used for storing solar charge in solar systems. The principles and operation of lead-acid batteries are outlined and types of commercially available batteries are described. The problems caused by the deep discharge and overcharging of batteries are discussed in detail. The practice of battery management and maintenance is also discussed and the tools needed to maintain batteries are detailed.

Energy Storage

Solar cell modules generate electricity only when the sun is shining. They do not store energy. A few applications, such as attic fans or pumps, do not need batteries because the appliances only need to work when the sun is shining. For example, a solar-pump lifts water into a storage tank and water flows down from the tank when needed. Attic fans only need to blow air out of the attic when it is hot (i.e. when the sun is shining).

However, for other applications we need electricity when the sun is not shining. Electric charge generated during the day must be stored to be available at night or when it is cloudy.

The most obvious answer to this problem is to use batteries (sometimes called accumulators), which chemically store electric charge. Other proposed solutions to energy storage include flywheels (which store energy in rotating wheel-like masses), compressed air, fuel cells and pumped water. Nevertheless, today, such solutions are neither practical nor economical for small systems. In fact, virtually all off-grid lighting and refrigeration systems use some type of battery to store their harvested solar energy.

Stated simply, a battery is like a tank for electric energy. The solar array produces an electric charge as long as the sun is shining. The charge travels through wires into the battery where it is converted to stored chemical energy. Over the course of several days, a battery may 'fill' with stored energy like a water tank 'fills' with water collected from rooftop gutters.

It is impossible to remove more energy from the battery than is put in by charging. As is the case with the tap of a water tank left open, if an appliance is left on by accident, electricity will drain from the battery. Batteries operate like energy bank accounts. From the bank, as with batteries, you can never get something for nothing. You can not take out more energy from a battery than was put in. Also, like a bank account, the long-term benefits from a battery are greatest when a large amount of energy is kept in the battery.

General Points to Keep in Mind about Batteries

Solar system planners need to keep in mind several points about batteries. As the battery is often the weak link of a solar power system, it needs to be selected carefully. Poorly functioning batteries reduce system performance and can damage appliances such as lamps and television sets.

- Batteries make up the largest component cost over the lifetime of a solar system. Good batteries are expensive, but worth the investment if you can afford them. Try to choose the right type of battery for your system.
- Try to ensure batteries get a full charge regularly (and, for flooded-cells batteries, are occasionally equalized).
- Batteries wear out. No matter the type, batteries eventually wear out and need to be replaced. Find out how long your battery is expected to last from the supplier. Plan for the replacement of your battery.
- Most batteries need to be maintained. Keep acid/electrolyte levels topped up and all surfaces clean.
- Do not mix battery types in your battery bank. All your batteries should be of the same type and manufacturer, and about the same age. Old or poorly performing batteries decrease the performance of those to which they are connected.

Some final words of warning about batteries:

- Every make of battery has different characteristics. So be careful when selecting and try to learn about the type you buy. This chapter is a general guide only – battery specifications sheets and manuals should always be referred to for definitive information on battery characteristics, installation, commissioning, maintenance and performance.
- Terminology is not always consistent. Some dealers call their batteries 'solar batteries', but this does not necessarily mean they are 'true' solar batteries!

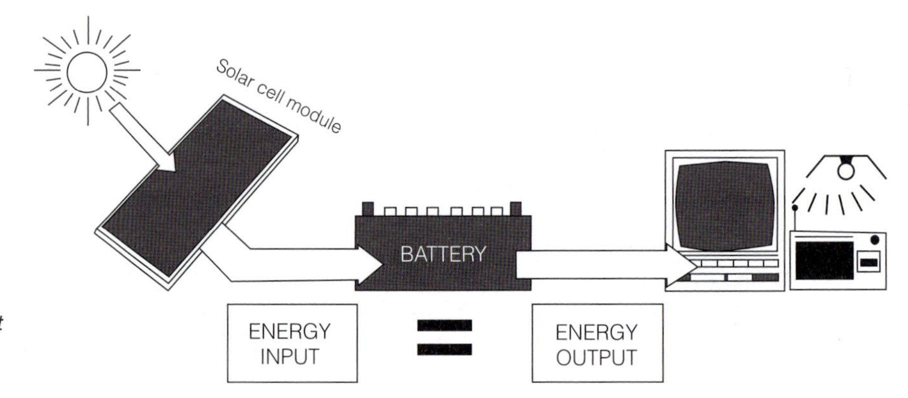

Figure 4.1 *Energy input from the solar array must balance the energy use of the load*

Battery Principles and Operation

Batteries are groups of 'electrochemical cells' – devices that convert chemical energy into electrical energy – connected in series. (Battery cells should not be confused with solar cells, which operate according to completely different principles.) Battery cells are composed of two 'electrodes' (also called plates) immersed in an 'electrolyte' solution. When a circuit is formed between the electrodes, a current flows. This current is caused by reversible chemical reactions between the electrodes and the electrolyte within the cell.

Some cells can only be used once – these are called 'primary batteries' (i.e. dry cells). Other batteries can be recharged over and over again; these are called 'secondary batteries' (or accumulators). Because ordinary dry cells cannot be recharged, this chapter is concerned with rechargeable secondary batteries only.

As a battery is charged, electric energy is stored as chemical energy within the cells. When the battery is being discharged (i.e. when it is connected in circuit with a load), stored chemical energy is being removed from the battery and converted to electrical energy.

The most common types of rechargeable battery systems on the world market today are lead-acid, lithium ion, nickel metal hydride and nickel cadmium. Lead-acid batteries are far more readily available, cost-effective and suitable for all but the smallest solar electric power systems. The latter three types of batteries are usually used for small electric appliances such as laptop computers, cell-phones, radios and lanterns.

As indicated by its name, the lead-acid battery operates on the basis of chemical reactions between a positive lead dioxide plate (PbO_2), a negative lead plate (Pb) and an electrolyte composed of sulphuric acid (H_2SO_4) with water (H_2O). When a battery is being charged, lead dioxide accumulates on the positive plate, spongy lead accumulates on the negative plate and the relative amount of sulphuric acid in the electrolyte increases. When the battery is being discharged, lead sulphate ($PbSO_2$) accumulates on the negative plate and the

Box 4.1 Chemical equation for lead-acid battery charge and discharge

← Charge

$$Pb + PbO_2 + 2H_2SO_4 \rightleftharpoons 2PbSO_4 + 2H_2O$$

Discharge →

Note that this is a reversible reaction. 'Charge' occurs when the reaction is moving to the left (i.e. lead and sulphuric acid are produced) and 'discharge' occurs when the reaction is moving to the right (i.e. lead sulphate and water are produced).

relative amount of water in the electrolyte increases (see chemical equation in Box 4.1). Each cell in a lead-acid battery has a voltage of about 2.1V when fully charged. This means that a fully charged 12V lead-acid battery has a voltage of 12.6V when not connected to anything. That voltage increases when it is being charged by a solar module and decreases when it is being discharged by a load.

Nickel Metal Hydride, Nickel Cadmium and Lithium Ion Batteries

These batteries are similar to each other (and dissimilar to common lead-acid batteries) in that they are mostly sealed, portable, require no maintenance and are used to power small consumer electronic devices (such small batteries are also known as 'pocket plates'). As with all batteries, they operate on the basis of chemical reactions between positive and negative electrodes in an electrolyte. Each fully charged nicad (the abbreviation for nickel cadmium) or metal hydride cell has a voltage of about 1.3V (lithium ion cell voltages range between 3.3 and 4.0V/cell).

In general, these types of batteries are more expensive per unit of storage than lead-acid batteries – this is the reason most PV installations choose lead-acid batteries. However, they have some advantages that should be considered by very small system planners:

- They are available in lighter and smaller sizes. When packaged in common dry cell sizes (see Figure 4.2), they can be used to power portable appliances such as radios, cassette players, computers, electric drills, cell phones and torches.
- They require less maintenance than most lead-acid batteries, an important consideration at sites where the system maintenance is a problem.
- They often have a longer life compared to most lead-acid types.
- Nicads and metal hydride cells can be completely discharged without damage to the cells, and they can be left for long periods in a low state of charge. (Lithium ion batteries are damaged by deep discharge.) They still require charge regulators though!
- Nicads and metal hydride batteries can be operated over a wider range of temperatures than lead-acid batteries. Lithium ion batteries perform best in cool temperatures (10–20°C or 50–70°F) and are damaged by heat.
- Lithium ion batteries do not suffer from the 'memory effect' common in nicads. However, they may explode if mistreated or if

Figure 4.2 *Rechargeable lithium ion and nickel metal hydride battery cells*

overcharged. Much research is going on with lithium ion batteries to improve costs and performance, and they are expected to be more widely used for PV and electric cars.

- Many solar system users keep a few nickel metal hydride cells for use in torches, lanterns, radios and consumer electronic devices, saving on the cost, waste and inconvenience of dry cells.

Lead-Acid Batteries

Lead-acid batteries are the most readily available solution to the problem of storing PV charge. Therefore, unless noted, discussions that follow in the following sections concern lead-acid batteries only.

There are a number of types of lead-acid batteries (as explained below), but all fall into two general categories: 'deep discharge' and 'shallow discharge'. Deep discharge batteries are preferred for solar electric systems because most of their stored energy can be delivered without causing damage to the cells or shortening their life. Shallow discharge batteries, made for automotive purposes, are designed to supply a large amount of power for a short duration. Taking too much energy out of these batteries before recharging them is likely to damage the plates inside. If chosen for solar electric systems, shallow discharge batteries should be managed very carefully and never deeply discharged.

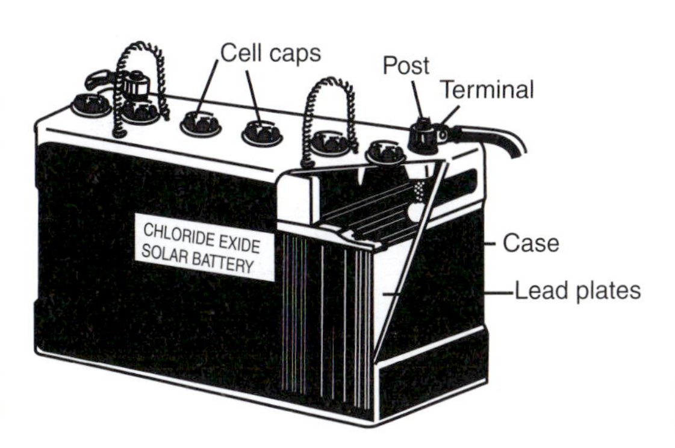

Figure 4.3 *Parts of a lead-acid battery*

Box 4.2 Lead-acid battery hazards

- Lead-acid batteries contain corrosive sulphuric acid. If spilled, sulphuric acid will burn the skin or eyes (and potentially cause blindness). It will burn holes through clothes and furniture and it will damage cement floors.
- If acid is splashed into the eyes, rinse each eye continuously for ten minutes with clean water; then take the affected person to the hospital for further treatment.
- If acid gets splashed on skin, wash immediately with plenty of water.
- If acid spills on the floor, pour bicarbonate of soda on the acid and rinse the floor with plenty of water.
- Batteries give off explosive hydrogen gas when they are being charged. This gas must be vented away from the battery to prevent explosions. Do not smoke or carry open flames in battery storage rooms.
- Batteries contain a large amount of energy that can cause an explosion or fire if the battery terminals are accidentally 'short-circuited' (i.e. when the positive and negative terminals are accidentally connected by a metal conductor, releasing all the energy stored in the battery more or less instantaneously). Make sure no metal objects can accidentally be placed across the terminals. Rings

and jewellery should be removed and tools placed on the ground when working on batteries so that they do not fall on battery terminals and create a short circuit. Tools should also be insulated. It is best to place batteries in a vented box in a room to which children and pets do not have access.
- When commissioning batteries they often need to be filled with sulphuric acid. Goggles and gloves should be worn and an acid pump used if possible, rather than a funnel.
- Working on large battery banks and in battery rooms is extremely hazardous and should only be carried out by appropriately trained persons.
- Battery installation and commissioning manuals should be read and followed.
- Batteries contain toxic materials; they need to be disposed of correctly and/or recycled.
- In larger battery rooms two compartments are required – one for the batteries and one where switchgear, fuses and inverters are installed. This is because electric gear can spark (e.g. switches and relays) and ignite explosive gases.

Rated Storage Capacity

The amount of energy that a battery can store is called its *capacity*. A water tank, for example, with a capacity of 8000 litres can hold at *most* 8000 litres. Similarly, a battery can only store a fixed amount of electrical energy.

Battery capacity, typically marked on the casing by the manufacturer, is measured in amp-hours (Ah). This indicates the amount of energy that can be drawn from the battery before it is completely discharged. The concept of the amp-hour is based on the notion of an ideal battery of, say, 100Ah, which ideally gives a current of 1 amp for 100 hours, 2 amps for 50 hours, and 4 amps for 25 hours (i.e. 1 amp times 100 hours, or 2 amps times 50 hours, or 4 amps times 25 hours equals 100 amp-hours). However, the rate at which a battery is discharged affects the capacity of the battery; for example, a battery discharged at a rate of 1 amp might provide that current for 100 hours, giving it a capacity of 100Ah at that rate of discharge; but the same battery discharged at the rate of 4 amps might only deliver that current for 20 hours thus giving it a real capacity of 80Ah.

On battery datasheets this is indicated by C-rates: C100 indicates the capacity of a battery being discharged over 100 hours (at the rate of 1 amp in the above example), C20 indicates the capacity of a battery being discharged over 20 hours (at the rate of 4 amps in the above example). Thus, the battery in the example has the following capacities: 100Ah at C100 and 80Ah at C20. So when choosing a battery you should be aware of the C-rate to which the rated storage capacity relates. You should also be aware that in a real working system the rated storage capacity is always a general guideline and not an exact measurement of the battery's energy storage parameters. In most systems, however, batteries will operate somewhere between C100 and C20. Capacity changes also with a battery's age and condition, the temperature and the rate at which power is drawn from it. If current is drawn from the battery at a high rate, its capacity is reduced. The capacity of lead-acid batteries is reduced with decreasing temperature – a typical battery will hold about 20 per cent less charge at 0°C (32°F) than one at 40°C (104°F).

Note that amp-hours are not a measure of energy – to convert amp-hours to watt-hours, multiply by the battery voltage.

Charge and Discharge

'Charge current' is the electric current supplied to and stored in a battery. As a water tank will take more or less time to fill depending on the rate at which water enters it, the amount of time required to completely charge a battery depends upon the rate of the current at which it is being charged.

Batteries can be charged by a variety of power sources:

- by solar cell modules (preferably using a solar charge controller);
- by grid power (using a battery charger);
- by wind power (using a wind-turbine charge controller);
- by petroleum generators (using a battery charger); or
- by automobile engines (using an alternator).

If the charge received (Q in amp-hours) is multiplied by the battery voltage, then the energy supplied to the battery will be given in watt-hours.

Of course, some energy is always lost in the charging and discharging process as heat. Depending on the type of battery and its age, the energy lost is between 10 and 30 per cent for lead-acid batteries. Very old batteries have even lower efficiencies.

Low currents (i.e. 3–5 per cent of the battery capacity) are best for charging batteries. Batteries should not normally be charged at currents that are higher than one-tenth of their rated capacity. Thus a 70Ah battery should never be charged at a current of more than 7 amps. When batteries are charged at high currents, electrolyte is lost rapidly through gassing; also, cell structure may be damaged. In a 12 volt system, the solar charge from a 12V 50Wp module does not get much higher than 3 amps, which is well-suited for charging a 70 or 100Ah battery.

'Discharge' is the state when battery energy is being consumed by a connected load (i.e. appliances). The discharge current is the rate at which current is drawn from the battery. The amount of energy removed from a battery over a period of time can be calculated (as charging energy was determined above) by multiplying the discharge current by the amount of time the load is used. For example, a lamp drawing 1.2 amps for 4 hours uses 4.8 amp-hours of energy from the battery (1.2A × 4 hours = 4.8Ah).

Box 4.3 Calculating the charge of a battery

The amount of charge a battery has received ('Q', in amp-hours) can be approximately determined by multiplying the charging current ('I', in amps) by the amount of time the current has been left on ('T', in hours):

$$Q = I \times T$$

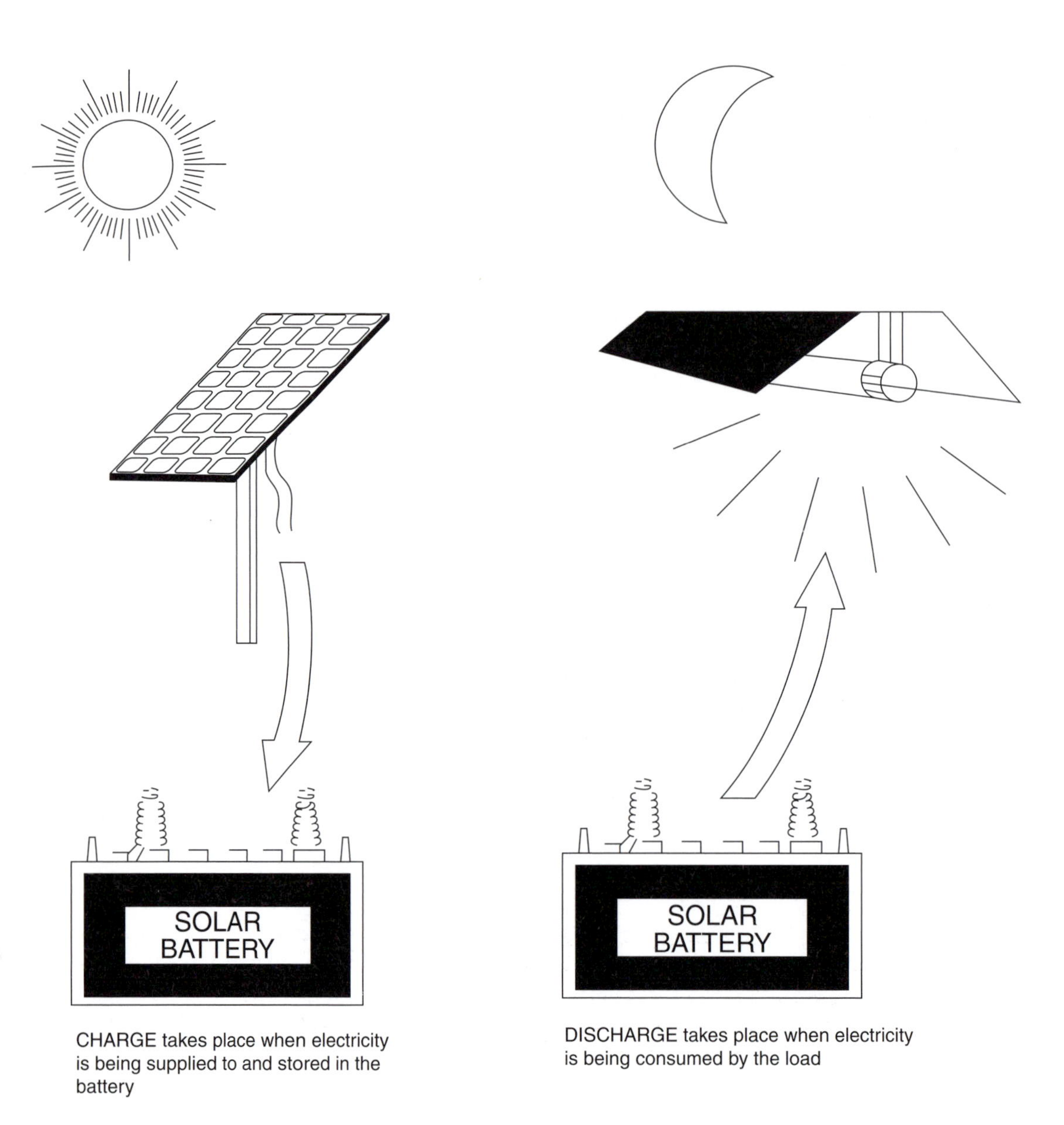

Figure 4.4 *Charge and discharge*

State of Charge

Just as one needs to monitor the amount of fuel left in a car's petrol tank, one needs to keep track of how much energy remains in the battery. The 'state of charge' (SoC) is a measure of the energy remaining in the battery. It tells you whether a battery is fully charged, half-charged or completely discharged. The cells of a fully charged battery have a 100 per cent state of charge, while those of a battery with one-quarter of its capacity removed are at a 75 per cent state of charge.

Measuring State of Charge

The state of charge of a PV-system battery is checked to determine whether the battery has been discharged too much and to determine the condition of the individual cells. Ideally, solar electric system owners should check the battery (or battery bank) state of charge at least once every two months (many charge controllers will indicate state of charge, see Chapter 5). Heavily used small systems should have the battery state of charge checked once per week during cloudy weather.

With lead-acid batteries (but not with most other types of batteries) the voltage of an open circuit cell varies with the battery state of charge. Thus, it is possible to roughly determine state of charge using a 'multimeter' (which measures the voltage of the battery and cells). Similarly, the density of a lead-acid battery electrolyte changes depending on the state of charge. A 'hydrometer' measures the thickness of sulphuric acid in each cell.

Good hydrometers are far more accurate for measuring state of charge than multimeters and can also to be used to detect dead cells. Nevertheless, multimeters are more convenient, versatile, less hazardous and easier to use. Multimeters are also necessary for other tasks (such as measuring module output) and, because hydrometers cannot be used on sealed batteries, they are often the only way of estimating a battery's state of charge. Owners of large systems (ones with several batteries) can also use them. Solar technicians need to be able to use both multimeters and hydrometers.

When measuring state of charge, always check the electrolyte level in each cell to make sure that it has not fallen too low due to gassing. The level should be well above the plates – on better quality batteries the recommended level is indicated by a mark.

Measuring state of charge with a multimeter

As the state of charge of a lead-acid battery decreases, its voltage also decreases and this can be measured with a multimeter set to measure DC volts (a multimeter has several measuring functions and can act as a voltmeter). A typical solar battery at 100 per cent state of charge has a voltage of about 12.6V. When discharged to 50 per cent SoC its voltage will be about 12.1V and when completely empty (or dead) its voltage will be about 11.5V or lower (see Table 4.1). The actual reading varies with the type of battery and the temperature – battery manufacturers should provide information about their battery state of charge parameters. A battery on charge (e.g. connected to a solar module during the day) will have a higher voltage and a battery connected to a load will have a lower voltage.

To measure a battery's state of charge with a voltmeter:

1. Disconnect the battery from the load and solar charge. If the battery was being charged (or discharged), wait at least 20 minutes to allow the cell voltages to stabilize before taking a measurement. If you measure right away, the reading will be inaccurate.
2. Connect the voltmeter's leads to the positive and negative terminals of the battery or cell. Read the voltage on the voltmeter and compare it to the reading on a state of charge table that is appropriate to your battery (see Table 4.1).

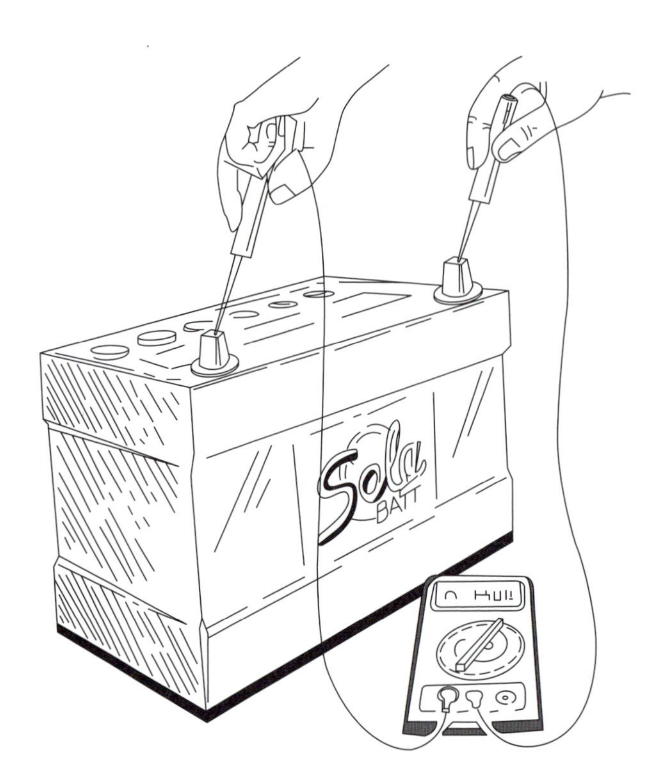

Figure 4.5 *Measuring state of charge of a battery with a digital multimeter*

Note that measurements of 12V batteries are the combined ratings of 6 cells, each of which is 2.1V at full charge). If 2V or 6V batteries are used in a system, it may be necessary to measure each of these.

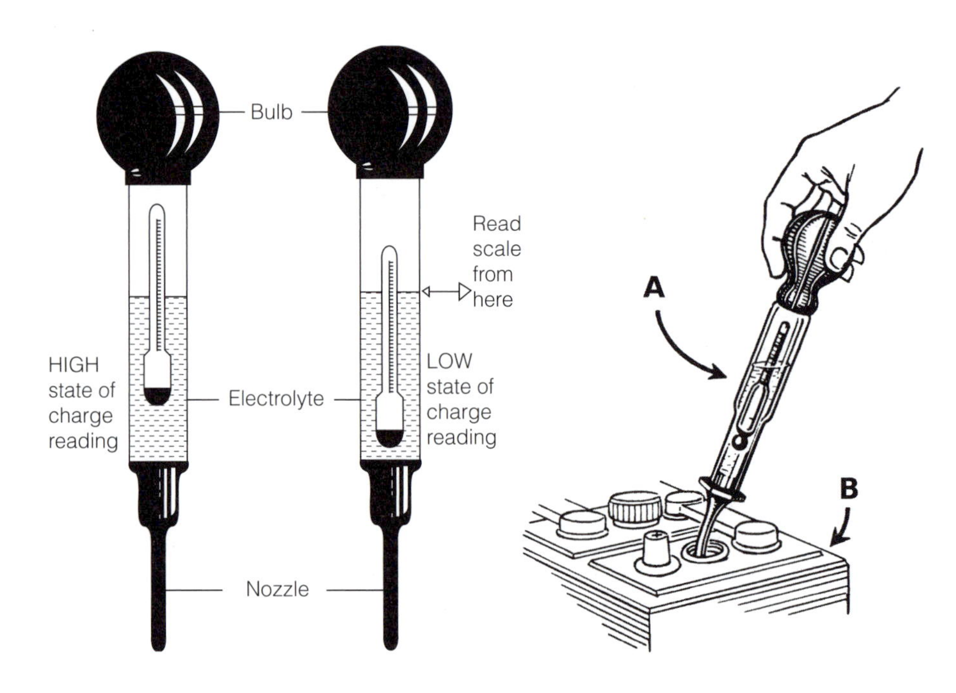

Figure 4.6 *Using hydrometers*

Table 4.1 *Measuring state of charge for a typical 12V modified SLI-type battery*

State of charge	12 Volt battery	Volts per cell	Specific gravity, grams/ litre at 25°C (77°F)
100%	12.7	2.12	1230
90%	12.5	2.08	1216
80%	12.42	2.07	1203
70%	12.32	2.05	1189
60%	12.20	2.03	1175
50%	12.06	2.01	1161
40%	11.9	1.98	1147
30%	11.75	1.96	1134
20%	11.58	1.93	1120
10%	11.31	1.89	n/a
0	10.5	1.75	n/a

Source: Manufacturer's data

Measuring state of charge with a hydrometer

Hydrometers measure the 'density' (this is also called the 'specific gravity') of the sulphuric acid electrolyte in each cell, which is directly related to the state of charge of the battery. As lead-acid batteries are discharged, the sulphuric acid within each cell is converted to water, which has a lower density than sulphuric acid. Thus, discharged electrolyte becomes less dense as the battery's state of charge decreases. Hydrometers contain a floating scale with specific gravity readings that measure this density (see Figure 4.6).

Be aware that if the battery acid has 'stratified' (see below) this will give a false hydrometer reading. Be extremely careful when using hydrometers; wear protective goggles and always have supplies of water at hand to deal with acid spillages (see Box 4.2), clean hydrometers thoroughly after use and never leave them lying around.

Use the hydrometer as follows:

1. Draw sulphuric acid up into the hydrometer from the battery cell by squeezing the bulb while the nozzle of the hydrometer is placed in the cell (see Figure 4.6).
2. The scale floats at a level that varies according to the density of the acid and the state of charge of the cell.
3. Read the specific gravity of the cell from the scale floating in the acid (see Figure 4.6). Some hydrometer scales do not give the specific gravity, but only indicate whether the battery is in a low, medium or high state of charge – scales giving numbers are preferable.
4. Consult a state of charge v. specific gravity table or graph to determine the SoC (i.e. see Table 4.1).

Cycle, Cycle Life, Depth of Discharge and Maximum Discharge Currents

Typically, batteries in PV systems are charged each day by the PV array and then discharged by the load each night (though this is not always the case in larger systems where some loads are also powered during the day). Each charge period together with the following discharge period is called a 'cycle'. For example, in one cycle a 100Ah battery might be charged up to 95 per cent state of charge (12.68V) during the day, and then discharged by lights and television to 75 per cent state of charge (12.42V) that evening.

The 'rated cycle life' of a battery (this should be specified by the manufacturer) is the number of cycles a battery is expected to last before its capacity drops to 80 per cent of its original rated capacity. Note that, in off-grid systems, this is typically the number of days the battery will last because each 'day' is more or less the same as one 'cycle'. Note also that the cycle life is determined by the average depth of discharge per cycle – a battery cycled at 30 per cent will last longer than a battery cycled at 70 per cent – as well as by average battery temperature. Figure 4.7 shows that a deeply cycled battery has a much shorter life than a shallow cycled battery. The actual cycle life of a battery is greatly shortened by such mistreatment as deep discharge (see below), high temperature and high discharge rates. A temperature difference of 10°C (18°F) can halve the cycle life.

'Depth of discharge' (DoD) is another term that manufacturers use to express how much batteries are discharged in a cycle before they are charged again. A battery at 20 per cent DoD is the same as an 80 per cent state of charge battery. A battery at 75 per cent DoD is at a 25 per cent state of charge. Shallow cycle batteries should not be discharged below 20 per cent DoD (80 per cent state of charge) on a regular basis. Even deep cycle batteries should not regularly be discharged below 60 per cent DoD (40 per cent state of charge). Remember, batteries last much longer when they are maintained in a high state of charge (see Figure 4.7).

A 'deep discharge cycle' is a cycle in which a battery is almost completely discharged. This typically occurs during long cloudy periods or when the load is much larger than the solar charge.

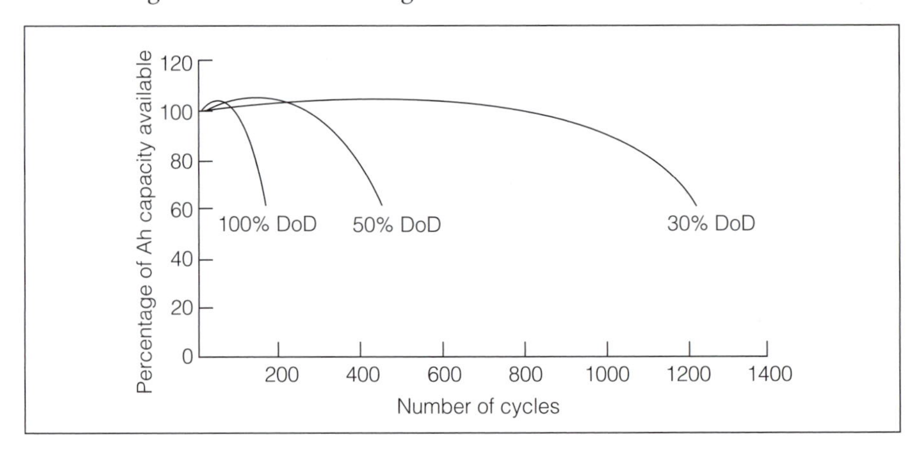

Figure 4.7 *Battery cycle life is greatly reduced with deep discharges*

Table 4.2 *Expected cycle life of two battery types at various DoDs*

Daily depth of discharge	Daily state of charge	Modified automotive battery	Deep discharge tubular plate solar battery
10% DoD	90%	750 cycles	10,000 cycles
50% DoD	50%	310 cycles	3000 cycles
80% DoD	20%	200 cycles	1200 cycles

Table 4.2 shows how, for two types of batteries, the rated cycle life is shortened by deep discharges.

The most common cause of failure in small off-grid PV systems is abuse of batteries by deep discharge during cloudy weather. Remember, when there is less sunshine, appliance use should be reduced. After a deep discharge, always let the battery recover by allowing it to reach a full state of charge before using it again. This will greatly increase the battery's life.

The 'maximum discharge current' of a battery refers to the maximum current that can be taken from a battery at any one time without significantly shortening battery life. This can be an issue especially when large inverters are connected to batteries. Battery specification sheets should give details, or the C-rates given for the battery can be used as a guide.

Self-discharge

If they are left standing uncharged, all batteries lose charge slowly by a process called 'self-discharge'. This occurs because of reactions within the cells of the battery (for example, cars left unused for several months often fail to start due to self-discharged batteries). Lead-antimony batteries have a much higher self-discharge rate (2–10 per cent per week) than lead-calcium batteries (1–5 per cent per month).

The rate at which batteries self-discharge depends on the temperature, the type of battery, their age and condition. As batteries get older, self-discharge rates increase. Also, dirty batteries (i.e. those with a high accumulation of acid mist and dirt on top) tend to have higher self-discharge rates, and high ambient temperatures increase the rate of self-discharge. Normally, new batteries do not discharge more than 5 per cent per month. In hot weather, old uncharged Starting, Lighting and Ignition (SLI) type lead-acid batteries lose 40 per cent or more of their capacity per month.

To avoid self-discharge:

- store the battery off the floor in a wooden battery box or non-metallic tray;
- keep the top surface of the battery clean;
- keep the terminals clean and greased.

Avoid storing a battery for a long time without charging it. Lead-acid batteries left in a low state of charge for long periods lose some of their capacity due to a permanent chemical change in the plates called 'sulphation'. If a battery is left in a low state of charge for over a month, it may not accept its rated charge capacity, or it may not accept charge at all. (Note that some types of gel cells can be left discharged for long periods without damage).

All batteries – and especially imported batteries that may have been stored in warehouses or customs for a long time – should be fully charged before putting them into use. This should be part of the commissioning procedure (see Chapter 9).

Stratification

Stratification occurs when electrolyte in flooded-cells separates into layers, with the heavier electrolyte 'sinking' to the bottom of the cell and the lower density electrolyte floating to the top of the cell. Not only does this give false hydrometer readings, it also decreases the performance of the cell. Avoid stratification by occasionally giving an equalizing charge to the battery – the intense bubbling of equalization mixes up the acid.

Overcharging and Charge Controllers

Batteries left connected directly to a solar module (without a charge controller) after they have been fully charged can become 'overcharged'. If, for example, a battery at 100 per cent state of charge is still directly connected to a solar module on a sunny day, the excess current will cause it to be overcharged. Occasional overcharging will not damage flooded-cell batteries; however, continued overcharging causes loss of electrolyte, damage to the plates and a shortened cycle life. Overcharging will quickly damage most gel cell batteries.

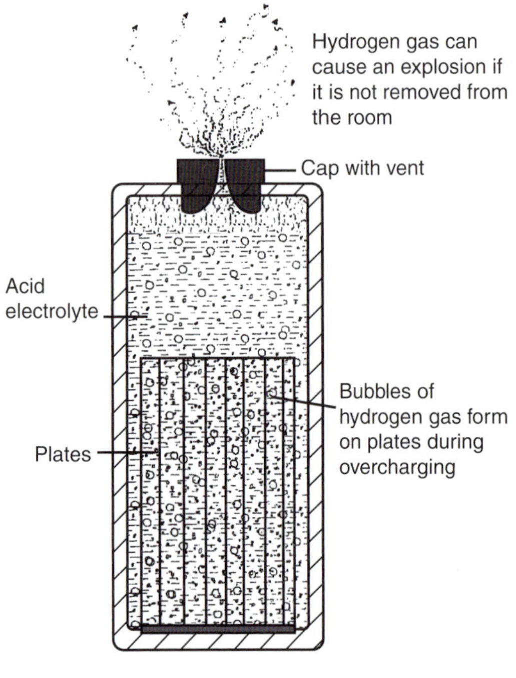

When a flooded-cell battery is overcharged, it loses electrolyte by 'gassing' (see Figure 4.8). Since a fully charged battery can no longer hold solar charge from the module, the charge causes a chemical reaction that converts water in the electrolyte to hydrogen. This gas escapes from the electrolyte as bubbles, causing two problems: firstly, the level of electrolyte in each cell goes down and distilled water must be added to replace it; secondly, explosive hydrogen gas is given off. To avoid the risk of explosion, this gas must be vented from the battery storage area.

When a sealed gel cell battery is overcharged, hydrogen bubbles will form in the gel material, and be unable to escape. This will damage or destroy the battery and can be dangerous!

In order to avoid damaging the battery by overcharging, always connect a 'charge controller' between the battery and the solar module (see Chapter 5). When the battery is fully charged, charge controllers reduce or stop the solar charge to the battery. Select the right type of charge controller for your battery.

Figure 4.8 *Gassing in a cell due to overcharging*

Types of Lead-Acid Batteries

Use the following summaries of common lead-acid battery types and their features to help choose the best battery possible for your system needs.

Grid plate batteries with fluid electrolyte (flooded/wet cells)

Automotive (SLI) batteries

Automotive (also called Starting, Lighting and Ignition or SLI) batteries are shallow discharge lead-acid batteries used mainly to start automotive engines. Because of their thin plates, they are damaged even when moderately discharged on a regular basis (i.e. below 80 per cent SoC). They are not good energy storage choices for PV systems. However, since they are locally manufactured, low-cost and widely available in many countries, they often end up powering lights and televisions in rural areas. For those installing small PV systems on limited budgets (especially in developing countries), 40–100 amp-hour automotive batteries are sometimes the only practical option.

If automotive batteries are chosen, special care should be taken to ensure they last as long as possible. They should not be used below 80 per cent state of charge (i.e. only 20 per cent of their capacity should be removed per cycle).

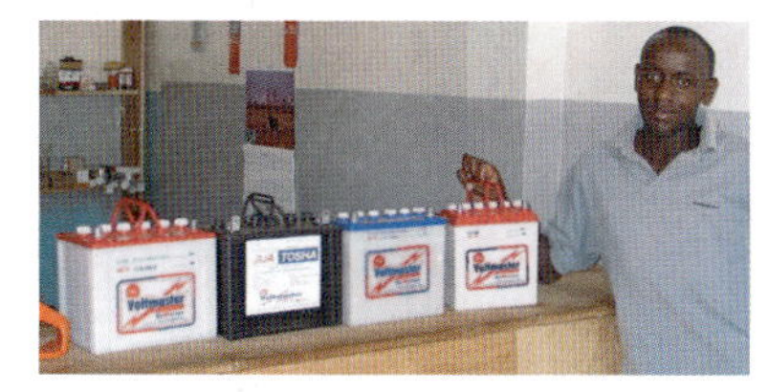

Figure 4.9 *Types of battery: Sunset flat plate solar battery (top left); local Kenyan SLI batteries, not generally recommended but often the only option (top right); Hoppecke 2 volt cell, flooded, tubular plates, deep cycle, recommended for larger systems (bottom left); Surrette 6 volt batteries, deep cycle (bottom right)*

Furthermore, during cloudy weather, they should be carefully maintained and taken for charging by alternative methods when necessary. If using automotive batteries, larger truck batteries, with 100–120Ah capacity, rather than low capacity 50Ah car batteries, are usually preferable. They are also only acceptable, if there is no alternative, in small systems with one or two batteries.

Improved automotive batteries (modified SLI)

Some battery companies manufacture 'improved' automotive batteries (sometimes labelled 'solar' or 'portable' batteries). With capacities between 70 and 200 amp-hours, these batteries have thicker plates, larger acid wells and, often, a handle for carrying the batteries. Modified SLI lead-antimony batteries are often used in small solar home systems (or as a grid-recharged power source for off-grid rural households). Although not the best choice for solar home systems, they are certainly better than standard automotive batteries. Common features include:

- A better cycle life than SLI batteries. Cycle life should be between 1000 to 2000 cycles (i.e. between 3–5 years) if the batteries are not discharged below 75 per cent SoC.
- A low self-discharge rate of 2–4 per cent per month.
- A large electrolyte reservoir to prevent damage from excess gassing and to minimize the need to add de-ionized water.
- Better tolerance of deep discharge than automotive batteries. Provided the battery gets a full charge regularly, it will have a good life under 30 per cent daily depth of discharge. Still, a daily discharge of 10 per cent gives the longest life (see Table 4.2).

Traction lead-acid batteries (also flat plate stationary batteries)

These types of batteries, developed for use in fork-lifts or golf-carts, are good for off-grid PV systems as they can tolerate deep discharges much better than automotive batteries and hence have a good cycle life. They have much thicker lead plates with a higher lead density than SLI batteries (which have 'spongy' lead). They are available in 2V, 6V or 12V sizes. However, because of their antimony content, they require frequent refilling from loss of electrolyte due to gassing.

Stand-by batteries

These batteries are used to power vital equipment (such as telephone exchanges) in the event of grid or generator failure. They have thick pure lead plates (without antimony addition) and are very heavy. They are not ordinarily designed for deep discharge and they are usually kept in a high state of charge because they are constantly on float charge. They do not have as long a cycle life as traction batteries. These batteries are occasionally seen in post offices or in relay station sites.

Lead calcium 'maintenance-free' batteries

These batteries are enclosed (but not completely sealed) and are sometimes sold for PV applications. Originally designed for automotive applications (i.e. Delco), their advantage is that they do not require electrolyte refill because of calcium added to their plates. However, the spongy calcium plates do not tolerate deep discharge well and can be damaged if left in a partial state of charge. Unless the system has enough power to maintain a very high state of charge, lead-calcium batteries should be avoided.

Valve-regulated lead-acid batteries (VRLA or sealed batteries)

If a battery is called 'maintenance-free' and is sealed (with no openings for adding acid) it is probably a valve-regulated battery. This means that, ordinarily, when gas is given off during charging, it is recombined into the battery as electrolyte (water). If the battery is overcharged, there is a safety valve that vents electrolyte and prevents dangerous build-up of gas pressure inside the battery. Note that VRLA batteries usually require special settings on charge controllers (see Chapter 5). VRLA batteries are much more expensive than ordinary lead-acid batteries and, because many use calcium in their plates to reduce gassing, they may be less tolerant of deep discharge.

Captive electrolyte (gel) batteries

Captive electrolyte batteries use sulphuric acid that has been turned into a gel form. Sealed at the factory, they do not leak or spill, so they are easily transported and require no maintenance. Some types can withstand deep discharges and have a good cycle life of 2 years when cycled to 50 per cent SoC and 3 years when cycled to 25 per cent SoC at 25°C (77°F). They have low self-discharge rates.

Note that many gel cell batteries are not made for deep discharging – always check the label/datasheet/manual. Captive electrolyte batteries have poor performance characteristics at high temperatures, so they should not be used in hot sites. Also, they should not be charged at high voltages or be heavily overcharged as this will cause a loss of electrolyte and may cause damage (or, in the worst case, an explosion). Small sealed gel cell batteries of 3–8Ah are commonly used in solar lanterns and consumer electronics.

Absorbed glass mat (AGM) lead-acid batteries

With AGM batteries, liquid sulphuric acid electrolyte is absorbed into glass fibre mats so they do not leak, even if cracked. Many AGM batteries are designed for stand-by 'float' applications, not deep discharging. AGM are often a good choice of batteries for off-grid solar PV systems, but, like gel batteries, they are also quite expensive. Some types are able to recombine gases, so they can tolerate overcharging and do not require special charge controller settings. Furthermore, they are extremely rugged and can withstand vibrations. AGM batteries tend to have a shorter cycle life than gel cell batteries, especially when deep discharged regularly.

Tubular plate batteries/OPzS or OPzV (wet or gel cells)

Tubular plate batteries, also called OPzS (liquid electrolyte) or OPzV (gel) batteries, are made especially for off-grid and solar electric applications and have excellent deep discharge characteristics. The positive plates in tubular cells are made of rods protected in a 'tubular' sleeve – not a flat plate – which gives them an exceptionally long cycle life. They often come in transparent cases that allow easy viewing of the electrolyte level and are sold in 2–6 volt sizes. They are among the most expensive batteries for off-grid installations and are ordinarily used for large installations. Tubular cells are susceptible to stratification and sulphation.

Choosing the Best Battery

Do not select battery banks for PV systems as an afterthought – battery life and performance will affect the overall performance and cost of the system. Before buying, find out what batteries are locally available on the market – both locally manufactured and imported. Because of their very short life-spans in off-grid PV systems, avoid automotive-type (SLI) batteries (but if SLI batteries are used, use truck rather than car batteries and make sure they are large enough).

When choosing your battery take the following things into consideration:

- Price. At the end of the day, a decision must be made on the cost of equipment. Some very good batteries are available for high prices – the buyer must decide what is important based on the particular needs of the system. In general, it is wise to spend more on batteries in larger systems. On the other hand, many small household solar PV systems in Africa and southeast Asia do quite well with low-cost modified automotive batteries.
- Capacity. Choosing the right size battery for your needs is important. Chapter 8 discusses how to size and choose your solar battery.
- Cycle life. In general, the better the battery, the longer the cycle life. Remember that batteries with lead-calcium plates have a lower tolerance for deep discharges. If your system is likely to be deeply discharged regularly choose lead-antimony plates instead.
- Replacement and availability. It is wise to choose a battery that is locally available. When the time comes to replace the set, it is more likely that you will be able to find a similar type. Shipping batteries across the world is expensive, time-consuming and energy intensive.
- Maintenance. Choose a battery that meets your maintenance needs and capabilities. If batteries are unlikely to be maintained, consider spending a little bit more on 'maintenance-free' units. Remember, no battery is truly 'maintenance-free' because the system must be managed or it will not work well!
- Size. Be aware of the physical size of the battery and where it is going to be placed. Make sure you use a safe area to enclose or house it.
- Mobility. If your battery is going to be moved around (not recommended but sometimes necessary) choose a type that will not be damaged by vibration and tilting.

Commissioning, Managing and Maintaining Batteries

When installing a solar system, make sure that the batteries are properly commissioned. Always make sure that the battery gets a full charge before it is put into use. Some gel cell batteries come dry-charged (meaning that they have a charge already). Lead-acid batteries that are filled on-site will only be partially charged. Always follow the manufacturer's commissioning instructions when completing your installation (see Chapter 9 for more information).

Depending on the type, batteries will last between two and ten years or longer if they are properly installed, commissioned, maintained and managed. They should be located in well-ventilated rooms (i.e. where air can circulate).

Tasks involved in maintaining and managing batteries include:

- Regular checking of state of charge to ensure that the battery is performing well; keeping state of charge records may help to detect when a battery is getting too old to use or when a cell has gone bad.
- For flooded-cell batteries, checking electrolyte levels in each cell; replacing electrolyte lost during gassing with de-ionized water. The plates should always be below the level of the electrolyte to avoid damage to the battery. De-ionized water, available at battery shops and garages, is used instead of tap water because it does not contain any impurities that could damage the cells. Never add tap water or acid to batteries.
- Cleaning the top of the battery. This avoids high rates of self-discharge caused by electrical conduction through acid mist accumulating on top of the battery.
- Cleaning terminals and contacts. Cleaning the terminals ensures a good electrical contact with the solar array and load. Application of petroleum jelly or grease to the terminals prevents them from becoming corroded.
- Giving the battery equalizing charges to mix up the electrolyte four to six times a year. Equalizing charges are charges well above the normal 'full' charge which cause the electrolyte in the cells to bubble and get mixed up and reduce risk of stratification or sulphation. Preferably, these charges should be done in the cloudy season or when the solar radiation is low. Note that some charge controllers automatically complete an equalization charge on a regular basis (see Chapter 5).

Replacing Batteries

Depending on the type and the way it is treated, a battery in a solar lighting system should last between two and ten years. However, at the end of its cycle life – when it will no longer hold a charge – a battery is dead and it needs to be replaced. With lead-acid batteries, performance decreases rapidly after the battery has aged to a point where its capacity is less than 80 per cent of the original rated capacity.

Typically, one or more cells will fail before the others, so it is a good idea to occasionally check for bad cells in old batteries. Do not waste your time replacing the cells of old batteries – this is usually futile, because they are likely to fail again. Battery manufacturers buy the cases of old batteries for recycling.

Replace entire sets of batteries at the same time. If a system has more than one battery in parallel, the batteries should be of the same age and condition. Putting a new battery in parallel together with an old battery will prevent the new battery from getting fully charged.

System owners should make a savings plan for the replacement of batteries. Batteries contain toxic chemicals, so they should be disposed of properly when they complete their life. In many countries, old batteries are purchased by scrap dealers or by battery companies themselves so that their lead and casings can be recycled. Always enquire about recycling and disposal of old batteries from dealers when purchasing new ones.

Bad Cells

Sometimes, a battery may have a bad cell. This means that, although the other cells in the battery are still working, one cell has stopped functioning properly (possibly due to a short circuit in the plates). When a battery's voltage is low (i.e. 10.5V or less), but the state of charge in most of its cells is high, the battery probably has a bad cell. To check for a bad cell, measure the state of charge of each of the cells in the battery individually using a hydrometer or voltmeter.

It may be possible to rebuild bad cells; certain artisans have set up businesses charging batteries and repairing bad cells. However, you should first check the work of such artisans before paying money to have a battery repaired. Quite often with old batteries, the repair job will only last several weeks before another cell goes bad in the battery. It is usually more economical to purchase a new battery rather than repair a single bad cell.

Alternatives to Solar Charging

Systems should be designed so that the solar array is large enough to keep batteries in a high state of charge even in cloudy weather. However, because of the high expense of modules, and because of the lengthy cloudy periods in some areas, small PV systems sometimes do not receive enough solar charge to meet the load requirements. In such cases, users should protect the battery from deep discharge by charging by an alternative method:

- Charging with a battery charger: the battery can be taken to a town for charging. In most small towns, there are petrol stations or businesses that operate mains-powered battery chargers that will top batteries up for a small fee.
- Charging from a car's alternator: if using automotive batteries in a system, the batteries can also be charged from the alternator of a car. This is done by replacing the battery in the car with the solar battery and either running the engine when the car is stationary for enough time to bring the battery to a high state of charge, or driving the car on a journey with the solar system battery in the place of the other battery.
- Use of other charging systems such as a genset (generator set) or wind-powered: in larger systems (>1000Wp) multiple charging systems are sometimes used. It may be desirable to combine a diesel generator or a wind generator with the PV array to make sure that the system has enough power in low radiation periods. See Chapter 11 for an example of a hybrid system.

5

Charge Controllers, Inverters and Load Management

This chapter discusses 'charge controllers' which are devices used to manage the energy flow in solar electric systems. The work of charge controllers and associated components is described, including overcharge protection, deep discharge protection, system power management and user alerts. The choice of charge controllers is briefly described. Management of small systems without charge controllers is also considered. This chapter also contains a short section on inverters for off-grid PV systems.

Why Use Charge Controllers?

As explained in Chapter 4, the success of any off-grid PV system depends, to a large extent, on the long-term performance of the batteries. For a system to operate well and have a long lifetime, the batteries must be charged properly and kept in a high state of charge. Over several months, the energy entering the batteries during the day (i.e. the solar charge) must be roughly equivalent to the energy being discharged from the batteries at night by the load.

Any off-grid PV system must be managed so that:

- batteries are not damaged by deep discharges from over-use of the load;
- batteries are not damaged through overcharging from the modules.

Solar electric systems use charge controllers (also called charge regulators) to manage the electrical power produced by the modules, to protect the batteries and to act as a connection point for all the system components (in systems without inverters).

The charge controller has a number of primary functions. Firstly, it provides a central point for connecting the load, the module and the battery. Secondly, it manages the system so that the optimum charge is provided to the batteries. Thirdly, it ensures that components (especially batteries and lights) are protected from damage due to overcharge, deep discharge and changing voltage levels. Finally, it enables the end-user to monitor the system and identify potential system problems.

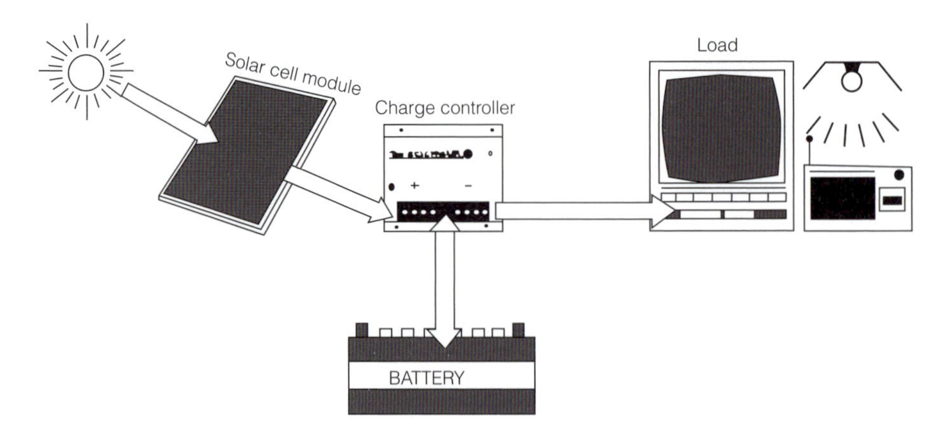

Figure 5.1 *The role of a charge controller in off-grid PV systems*

Low-voltage Disconnect

As mentioned previously, allowing batteries to deep discharge greatly reduces their cycle life. A charge controller's 'low-voltage disconnect' (LVD) feature continuously measures the state of charge of the battery. If the battery voltage drops below a certain level, the charge controller automatically disconnects 12V loads from the battery. This might happen during the cloudy season when the load is being used too much and when there is not enough charge to bring the battery state of charge up. Usually, a red LED lights up to notify the user that the battery has been disconnected. The controller will not reconnect the load until the battery voltage has returned to a suitable level (i.e. 12.3V) or, in the case of some controllers, until the user manually resets it.

On commercially available controllers, the low-voltage disconnect is commonly set at a point when the battery is below 40 per cent state of charge (i.e. between 11.1 and 11.7V). With some charge controllers, the level at which the controller cuts off the load can be adjusted. Before buying a charge controller, check the low disconnect voltage and be sure it meets your needs.

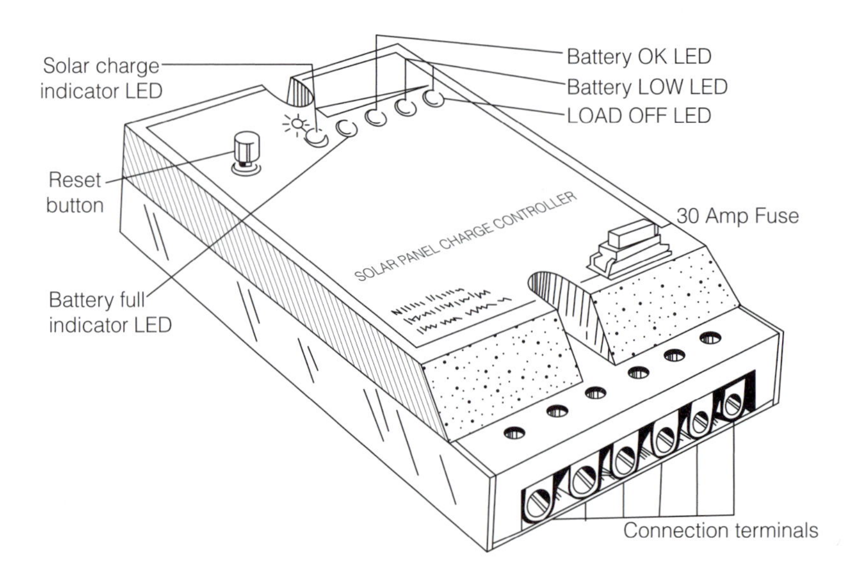

Figure 5.2 *LED indicators and features of a charge controller*

Many PV system users choose 240 or 110V AC appliances that are run directly through inverters. Inverter cables are almost always attached directly to the battery terminals (not the charge regulator). Thus, because inverters are connected directly to the battery, and not to the charge controller, low-voltage disconnects do not disconnect AC loads running through the inverter. Overuse of 240V AC loads through the inverter, therefore, can damage the battery over time by draining them below acceptable state of charge levels. Some inverters also have low-voltage disconnect features.

PV systems with inverters do not use a simple low-voltage disconnect to disconnect the inverter from the batteries when the batteries get low; this is because inverters can be damaged if they are disconnected from the battery when they are under load. In this case battery protective measures need to switch the inverter off first and then disconnect it from the battery. Most inverters do actually have low-voltage disconnects but these are designed to protect the inverter, not the battery! Battery monitors (see below) can also be used to manage systems. It is always essential in larger systems to have some type of device to alert users when the battery voltage is low (see below for a fuller discussion of this).

Some charge controllers will also measure the DC current being put into the battery by the solar module(s) and the current taken out of the battery to feed the DC loads. If there is an inverter in the system this function will not work properly because it takes current directly from the battery and effectively bypasses the charge controller. Such charge controllers will usually have a facility to switch this function off and rely on voltage measurements only – or will use a shunt to measure the current being fed to the inverter from the battery. Current measuring charge controllers also give a more accurate indication of battery state of charge, sometimes even giving it as a percentage, which is particularly useful for users.

Charge Controllers and Overcharge Protection

Charge controllers and the stages of battery charging

Different types of controllers change the rate at which they supply power to the battery over the course of charging. Common charging stages are bulk, absorption, float and equalization.

- Bulk Charge. During the 'bulk' stage of charging, which lasts until approximately 80 per cent state of charge, all of the power from the module is supplied directly into the battery.
- Absorption Charge. During the 'absorption' stage, the charge is gradually reduced until the batteries reach a 100 per cent state of charge.
- Float Charge. When the battery is full, the 'float' stage is reached and only enough power is supplied to maintain full charge of the batteries (also known as a 'trickle' charge).
- Equalization Charge. The 'equalization' stage, used only by some types of controllers (and utilized once a month or so), provides a short, high charge designed to reduce acid stratification, battery sulphation and to equalize voltages across battery cells.

Users of all sealed or gel batteries should make sure that the charge controller they choose is suitable or has a special setting for gel or VRLA cells (overcharging or equalizing VRLA/gel cell batteries will damage them and may be dangerous). Read the manual and make sure you understand the features of the charge controller before purchasing.

Types of charge controllers

Virtually all charge controllers contain a feature to prevent the array from overcharging the battery. The controller monitors the battery state of charge and reduces or terminates solar charge as the battery gets fully charged. Charge regulation technology has improved dramatically since the 1990s, so it is worth shopping around to find a modern unit. There are four general types of charge controllers (though they may share features), categorized by the method by which they regulate charge from the module to the battery as described below:

- 'Series-type controllers' rely on relays or electronic switches in series between the module and the battery to disconnect the module when the battery reaches a set voltage. When the battery state of charge goes down after the load is used, the controller resets the charger to turn 'on' during the next solar cycle. The simplest and lowest cost controllers, these are less common in modern PV systems.
- 'Shunt-type controllers' work in parallel between the array and battery. They gradually reduce power from the module to the battery as the battery reaches full charge, harmlessly short-circuiting it back through the module. In general, shunt-type controllers are low cost, fairly simple in design and well suited for small off-grid PV systems.
- 'Pulse-width modulation (PWM) controllers' send pulses of charge to the battery that vary depending on its state of charge. A battery with a low state of charge gets a 'wide' pulse (i.e. high charge) or the charge is on continuously. As the battery gets fully charged, the controller sends increasingly 'narrow' pulses of charge. A fully charged battery (in 'float' mode) just gets an occasional narrow pulse. The controller measures the state of charge and adjusts the pulse accordingly. (Both PWM and MPPT controllers use features of the series- or shunt-type controllers.)
- 'Maximum power point trackers' (MPPT) utilize DC to DC conversion electronics to 'track' the maximum power point of the module or array I-V curve (see Chapter 3). Remember, the maximum power point voltage of a module is often much higher than its battery-charging voltage, meaning that a 100Wp module might only be charging a battery at 75Wp at 14V (i.e. the charging voltage) – the maximum power point might be 16V. By keeping the charge voltage at the maximum power point, MPPT controllers gain an extra 10–35 per cent output from the array. MPPT trackers also can accept higher voltages from the module, meaning that they are able to use modules with a higher voltage (i.e. 24V or higher) and convert the power to 12V (or the rated battery voltage). MPPTs tend mostly to be used in large systems, where getting the most out of the array results in significant cost gains for the system.

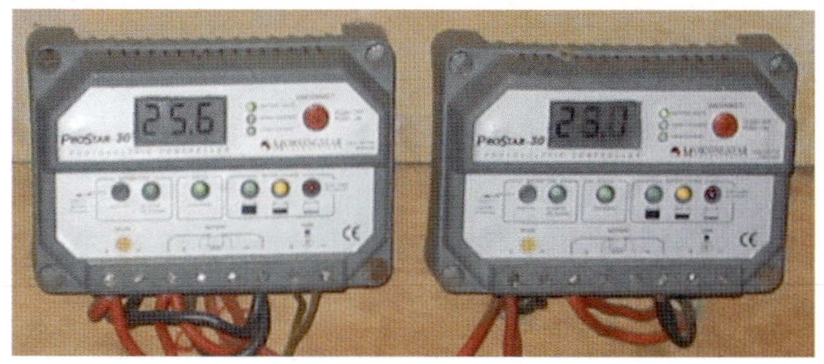

Figure 5.3 *Steca charge controller for a solar home system (top); Morning Star charge controllers for larger systems, with system indicators (middle); Phocos MPPT (bottom)*

Sources: Tobias Rothenwänder (top); Mark Hankins (middle, bottom)

Charge Controllers and System Connections

All off-grid PV systems should contain a distribution board or panel where major circuits are connected and where the charge controller is mounted. The charge controller provides a point where the battery, solar module and DC loads are connected. The panel (and some controllers) may incorporate switches and fuses to protect the equipment from damage by short circuits and overloads. As well, inverters, meters and other devices may be placed on this board (See Figure 5.4).

Circuit protection

Normally, protection devices (fuses or circuit breakers) should be installed between the battery and charge controller, between the load and the charge controller and between the inverter and charge controller. Most charge controllers contain internal fuses or circuit breakers that may serve one or more of these purposes, as explained below. Systems with inverters require separate protection devices on the AC circuit.

'Fuses' or 'miniature circuit breakers' (MCBs) protect the major circuits in a system from short circuits. Fuses are made of thin wires that melt when the rated current is exceeded, disconnecting the circuit. MCBs are small switches that automatically break the circuit when the rated current is exceeded. They can be switched back on when the wiring problem is corrected.

Some charge controllers contain fuses or MCBs inside their electronics. If they do not, you will have to install them on your system. As a minimum requirement, a main battery fuse should be installed (see Chapter 8 for information about selecting fuses).

'Surge protectors' protect system components and appliances against the rapid power increases expected when lightning strikes. Solar arrays are unlikely to survive direct lightning strikes. Nevertheless, surge protectors prevent damage to the system and appliances from nearby lightning strikes. Surge protectors operate by disconnecting the system when a very high current moves through the wire. Many charge regulators have inbuilt surge protectors.

Main switch and array switch

It is often necessary to control certain loads from the centrally located charge regulator using 'main circuit switches'. For example, in a school, classroom lights maybe switched on from the charge controller located in the office. This prevents misuse of lights by students in the classrooms. In a home, the lights can be turned off from the main circuit switch during the day to prevent draining of the battery by lights accidentally left on. It may be also be desirable to be able to switch the array on and off at the control panel.

Some charge regulators come with main switches. If the desired charge controller does not have a main switch or switch for the solar array, the system designer may want to include switches on the control panel.

Charge controller connection terminals

All charge regulators have positive and negative connection terminals for the solar array and battery. The size of the terminals depends on the rated size of

Figure 5.4 *Charge controllers, MPPTs, main switches, inverters, fuse boards and other accessories on wall-mounted panels*

Source: Mark Hankins

the controller. Most charge regulators have positive and negative connection terminals where 12V loads are attached.

Because many off-grid solar systems utilize inverters (see below), connection points for loads are not the only points where power is taken from the batteries. Inverters are almost always connected directly to the battery and not to the charge controller.

Other Charge Controller Features and Load Management Devices

Controllers may include other features to enhance the system's performance. It is also possible to purchase some of the devices mentioned below as separate components.

Temperature compensation

Some charge controllers have 'temperature compensation' features. Because batteries require a higher charging voltage at low temperatures, good quality charge controllers adjust the charging voltage depending on the temperature. Some controllers simply measure ambient temperature, but others have sensor cables which run from the controller to a thermometer mounted on the battery. The controller then automatically adjusts the charging voltage to the optimum charging level for a given temperature.

Load timers

'Load timers' are switches that connect and disconnect loads, making system management easier. They automatically turn loads on, limit the amount of time that the loads are kept on and prevent overuse of the battery. For example, in a school or clinic a load timer might switch classroom or ward lights so that they come on at sunset and stay on for a preset time before automatically being turned off. Another common use of load timers is for PV-powered billboard lights and security lights.

Battery monitors and amp-hour meters

'Battery monitors' measure the current and voltage of the load, batteries and modules, allowing end-users to determine exactly how their system has been performing. They provide useful information for users of systems including:

- voltage of the battery;
- array current;
- load current;
- daily (or monthly) total current input to the batteries (in amp-hours); and
- daily (or monthly) total current utilized by the load.

Good charge controllers incorporate a few or all of the above features and some even have remote monitors.

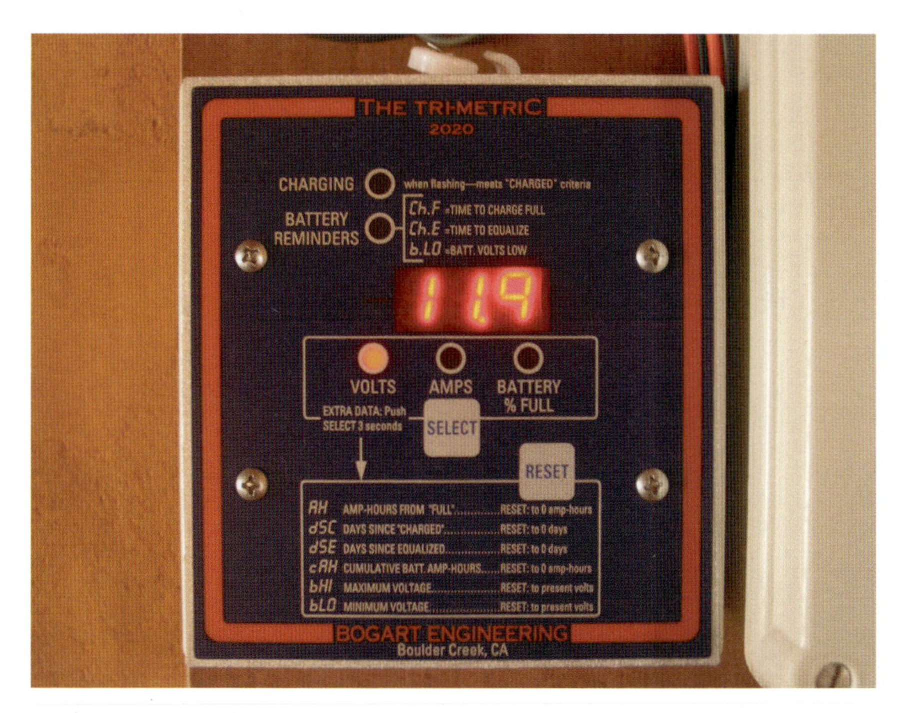

Figure 5.5 Battery monitor

Battery monitors are also available separately for mounting on the control panel (see Figure 5.5). For very small DC-only systems battery monitors are too expensive to be useful as separate devices. However, for larger systems they provide an excellent tool that enables managers to tell exactly what is happening in the system.

Choosing Charge Controllers

There is a large variety of charge regulators with a wide range of prices and features. When buying a regulator, always check its ratings and features to make sure you get what you need. Table 5.1 compares some commonly available charge controllers. (See also Chapter 12 for a list of websites for charge controllers and inverters.)

- Choose the right current (input and output) and voltage rating for your controller. Get assistance from a qualified designer if you need it. Charge controllers are rated according to the input current from the module, the output current to the loads (though this is very often the same as the input current) and the voltage of the battery. For example, a 10-amp unit will accept a maximum current of 10 amps from the module. The smallest available units are rated at 2 amps, which can handle 20Wp of power. Sizing of charge controllers is discussed in Chapter 8.
- Make sure that the charge controller has the features needed, including some type of alert to indicate the state of charge of the battery!

- Always read and keep the user manual. Make sure you understand each of your controller's features and, if you are the installer, how to install it.
- Make sure the charge controller is the right one for your battery. Remember to choose a unit with appropriate charging settings if you are using a gel or VRLA battery.
- Follow instructions when installing and operating charge regulators. Some units will not function properly, or can even be damaged or destroyed, if installed improperly (especially if connections are made in the wrong order, see Chapter 9).

Managing Systems without Controllers

Sometimes consumers, suppliers or designers of small PV systems omit charge controllers or choose not to install one in their system (this is usually for cost reasons). Without a controller, however, the battery is not protected against deep discharging. The result is often a battery that lasts for a very short time and a poor experience with solar energy.

Table 5.1 *Features of common charge controllers*

Feature	Blue Sky	Sundaya	Morningstar	Outback	Phocos	Steca
Model	Solar Boost 25	Apple	SunSaver	Flexmax	CX Series	Solsum
Type	MPPT	PWM	MPPT	MPPT	PWM Series	PWM Shunt
Current Rating	25A	5/10/15A	15A	60A, 80A	10-40A	5-10 A
Consumer Applications	Residential off-grid	SHS	SHS	Large PV systems >500W	SHS, residential off-grid	SHS
Battery Voltage	12V	12V	12/24V	12/24/36/48V	12/24V	12/24V
Load LVD	YES	YES	YES	No DC load connection	YES (5 algorithms)	YES (Units >6A)
Self Consumption	<1 W	N/A	35 mA	<1 W	<4 mA	4 mA
Charge Stages	Bulk, Absorption, Float	Bulk, Float	Bulk, Absorption, Float, Equalization	Bulk, Absorption, Float, Silent, Equalization	Bulk, Absorption, Float	Bulk, Float
Display & Alarms	LED • Battery SoC • Battery Disconnect	LED • Solar charge • Battery SoC • Alarm	LED • Solar charge • Battery SoC	LCD with 128 days data logger	LCD • Battery SoC • Energy Flows	LED • Solar charge • Battery SoC

Note: This is a representative introduction to a range of common charge regulators on the market. It is not intended to be exhaustive. For more information visit the company websites in chapter 12. All data from manufacturer.

This manual recommends that off-grid PV systems always utilize charge regulators. However, many consumers cannot afford to buy regulators. For example, many small PV systems in Africa or India use second-hand batteries that cost less than the price of a controller and omit charge controllers because they simply cannot afford the cost. For such customers, we recommend that, even if they cannot afford controllers, system owners should:

- learn to manage and maintain the systems (perhaps with a low-cost multimeter);
- install safety devices such as main switches and fuses;
- protect the battery through careful energy management, so that the investment is not ruined; and
- plan to include a charge controller on the system when it can be afforded.

The guidelines below suggest ways by which a system can be effectively managed without a controller.

- Check the battery state of charge regularly with a multimeter. Even without measuring, you can often tell when a battery's state of charge is getting low. For example, when battery voltage is low, fluorescent lamps take time to light up when turned on (or they may flicker). If the television picture gets dim and does not fill the whole screen or if music systems run slowly, this is a sign that the battery may be low!
- Calculate the approximate energy harvest of the module (see Chapter 8) and adjust energy use so that it is approximately the same as energy harvest. Reduce this during cloudy weather.
- When the battery state of charge is low, limit the use of appliances.
- Charge the battery by other means at least once every two months when it is being overused and more often during cloudy weather (see Chapter 4).

Inverters

'Inverters' (also called 'power conditioning units') convert lower voltage DC electricity into a higher voltage AC. They are often convenient for PV owners who wish to run appliances that require 110 or 230V AC (e.g. for colour televisions, laptops or music systems). Changing DC power into AC power is also called inverting DC to AC, hence the name inverters. Inverters are available in sizes ranging from 50 watts to thousands of watts. Check Chapter 12 for more sources of information about inverters for solar PV systems.

In the process of converting DC to AC, inverters use up energy. They are typically about 90 per cent or less efficient in converting power at full or near to full load (though there are some types that are even more efficient). At half-load or lower they will usually be less efficient. When planning large systems the energy loss in conversion must be included in calculations. Remember, inverters do not create energy – they actually use more energy and make the entire system less efficient.

There are two ways that inverters can be used in off-grid PV systems:

- In some systems, a small inverter is bought to power a single appliance. For example, small systems often have a 150W inverter attached to their system to power a colour TV. Usually, the appliance is plugged directly into the inverter. See Figure 5.6.
- In other types of systems, a larger inverter is used to power entire circuits. For example, one common arrangement is to have all lights in the system powered by 12V DC power and to use an inverter to supply AC power to sockets for appliances.

Choosing Inverters

Inverters should be carefully chosen to meet the needs of your solar PV system. Try to choose an efficient model made for PV systems! Remember, many inverters are not designed for solar use and perform poorly in PV systems because of poor efficiency or wave shape. It is always better to get a unit specially designed to be used in solar systems – check the manual or technical specifications (see Chapter 12 for a list of inverter manufacturers). Some inverters for off-grid PV systems come with integrated charge controllers (see below). Remember, never choose 'grid-tie' inverters for off-grid PV systems.

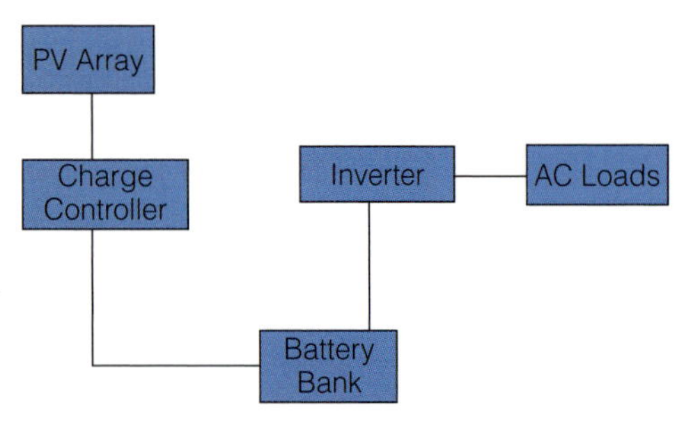

Figure 5.6 *Common inverter configuration*

Datasheets and instruction manuals should always be read carefully before buying (if the inverter you are considering does not have a datasheet with full information, this is a sign that it may be a poor quality unit).

Important factors to consider when selecting inverters include the following.

Type

Make sure the inverter is for off-grid systems, not a feed-in inverter for grid-connect systems. Be aware of the differences between an inverter and an inverter-charger. Most inverters do not have integrated charge controllers, but some do – consider the advantages and disadvantages of using one of these. Both of these are discussed below. In any case, read the inverter manual carefully (usually available on the internet) before purchasing.

Size

Inverters must be properly sized to handle the highest load that the system might require or the peak energy demand (see Chapters 8 and 11 for details on sizing inverters).

Voltage

Make sure your inverter matches the battery/controller voltage of your system and the voltage of your loads. Inverters are commonly available to accept 12V, 24V and, for very large systems, 48V DC. Depending on the part of the world you live in, inverters will be available to provide output power that matches the common AC power available in your region.

Efficiency

Usually inverters are between 60 and 95 per cent efficient. Efficiency varies greatly depending on the load and most inverters do not perform up to their efficiency rating all the time (they tend to be less efficient when loads are at less then 50 per cent of their capacity). Every inverter has an efficiency curve which will indicate its efficiency over a range of loads. Also, many poor quality low-cost inverters (and appliances) on the market are not designed for solar PV and are quite inefficient. Always check manufacturer's efficiency ratings and ask an expert if in doubt.

Wave form

This is a measure of how 'clean' the AC current output is. Inverters are classified into three types of output: 'sine-wave', 'modified sine-wave' and 'square-wave'. Square-wave inverters cost less, but may not be able to power some appliances (such as fluorescent lamps or videos); they may damage some equipment. Modified sine-wave inverters are better than square-wave and in general better for off-grid PV systems in terms of efficiency and overall performance. Sine-wave inverters, generally the most expensive type, have a wave shape that is very close to clean grid power. Use sine-wave inverters for appliances that are sensitive to wave form. To avoid disappointment, always make sure beforehand that the inverter you choose can handle all of the equipment you are running! This information is usually available in the manual. Inverters can interfere with radio and telecom equipment. If this is an issue, resolve it before buying an inverter.

Surge capability

This refers to inverter ability to supply sudden surges in power demand. For example, when a motor or refrigerator compressor turns on, it draws a large amount of power for a short time and the inverter must be able to deliver this surge. Surge capability of inverters should be provided in their datasheets and instruction manuals.

Phantom loads, stand-by mode and load detection

Even when appliances are turned off, many inverters still consume power from the battery (this is known as a phantom load). Consult the manufacturer's data for 'self-consumption' and tell system owners to switch the inverter off when it is not being used. Some inverters for solar systems shut themselves down when no load is turned on. They automatically detect when loads are in use, and switch themselves on. Such inverters use much less power than units that continually consume power even when no AC load is connected.

Using Inverters in Off-Grid PV Systems

Generally, in 12 or 24V DC systems one should try to use DC solar-designed appliances wherever possible. Efficient DC lights, refrigerators, music systems, computers, pumps, fans and even TVs are available for solar PV systems. If they are available use them because it will save energy losses through the inverter. Never connect thermal loads (cookers, coffee machines, blow driers, sterilizers, etc.) to inverters in small PV systems.

Just because you have an inverter does not mean you have more energy! Avoid using 'standard' inefficient AC appliances (desktop computers, refrigerators) just because you have AC power. Normal refrigerators or desktop computers will end up costing much more with the extra PV modules and batteries needed to power them. Remember, because the inverter takes power directly from the battery, and common off-the-shelf inverters will completely drain the battery without disconnecting the load, users need to be very careful about using inverter loads when the system is in a low state of charge. Inverters designed for use in off-grid PV systems are always preferable but users still need to be able to monitor battery state of charge.

Inverter-chargers in hybrid systems

Some types of inverters, called 'inverter-chargers', allow batteries to be charged from non-PV sources, such as diesel generators. Such inverters, usually rated 1000W or above, can charge the batteries when there is not enough solar power. Inverter-chargers are also used in power back-up systems in areas where there are frequent power-cuts; in this case they are connected to the grid rather than to a generator; PV can play a role in such systems, especially where the power-cuts are frequent and of long duration. See Chapter 11 for more information about hybrid systems and Chapter 12 for information about inverter-charger suppliers.

Inverters with charge controller features

Having AC circuits separate from the inverter – which feed power from the battery without low-voltage disconnect controls – greatly increase the risk of shortened battery life from deep discharge. Needing to buy both charge controllers and inverters adds considerable costs to PV systems.

To get around this problem in small off-grid PV systems, several manufacturers have introduced 'combined inverter-charge controllers' that serve as both inverters and charge regulators. With a single purchase, the needs of charge management, user alerts and AC power delivery are met, and the need for purchasing a separate inverter is eliminated. Table 5.2 includes information about such units which are available from Steca, Studer Innotec and other manufacturers (see also Chapter 12).

Inverters and inverter-chargers come with many different features such as back-up and uninterruptible power supply (UPS) options and can be quite sophisticated devices. One of the best ways to learn about them is to read inverter manuals.

Table 5.2 *Inverters for off-grid PV systems*

Feature	Victron	Outback	Studor	Steca	Solon	Morningstar
Model	Phoenix EasyPlus	FX2012T	AJ275	Solarix PI 550	Piccolo 150	SureSine
Application	Residential, marine	Residential	SHS, single appliance	Residential, cottage	SHS, single appliance	SHS, single appliance
Wave shape	Sine wave	Sine wave	Sine wave	Sine wave	Sine wave	Sine wave
Continuous Power Rating	1600VA	2000W	200W	450 VA	150W	300W
Surge Power Rating	3000W	4800VA Surge 2500VA (30 min)	275W	550 VA	400W	600W (10 min)
DC Input Voltage	12V	12V	12V	12V	12V	12V
Max Efficiency	N/A	90% (typical)	93%	93%	90%	92%
Stand-By Mode	N/A	YES (6W)	YES (0.3W)	YES	YES (170mA)	YES (55 mA)
Other features	70A Inverter/ Charger	60A Inverter/ Battery Charger	Can be supplied with integrated charge controller	Communicates with charge controllers to monitor battery SoC		

Note: This is a representative introduction to a range of common inverters on the market. It is not intended to be exhaustive. For more information visit the company websites in chapter 12. All data from manufacturer.

Voltage Converters

Many small DC appliances operate at a different voltage than the battery. These include radios, music systems, cell-phones or even laptop computers. For example, a radio may draw 6V DC while a PV system battery is likely to be 12V DC. 'Voltage converters' are used to step the current down to the proper voltage. This avoids damage to the appliance and improves the overall efficiency of the system.

Voltage converters are available in electric appliance stores and from speciality suppliers of solar electric equipment. Low-cost simple units are available made for radios and charging cell-phones. Units for cell-phones or laptops will save quite a bit of energy, but must be specially procured for the individual power need (see Chapter 12 for information on DC to DC converter suppliers). Make sure you get the right type of voltage converter for your appliance!

Figure 5.7 *Common sine wave inverters used in PV systems: Solon ASP Domino, 600W at 12V, 1000W at 24V (top left); Steca Fronius Solaris IP, 500W, version with integrated solar charge controller available (top right); Solon ASP Top Class, 1000W (middle left); Outback inverter with integrated battery charger for charging batteries from generator, 2000W (middle right); Studer-Innotec AJ Series, 400W (bottom left); Steca Solarix Sinus 500W, version with integrated solar charge controller available (bottom right).*

Figure 5.8 *Voltage converters*

Source: Steca

6
Lamps and Appliances

This chapter explains how to choose the best lamps and appliances for off-grid PV systems. Principles of efficient lighting are explained, including lumen output, efficacy and reflection. Information about incandescent, halogen, LED and fluorescent lamps (and their associated fixtures) is provided. Choice of lamp, depending on the purpose intended, is outlined. Important aspects of both DC and AC tools and appliances likely to be used in solar electric systems are presented.

Off-Grid PV Systems, System Voltage, Lights and Appliances

When putting together an off-grid PV system, the choice of power generation and storage equipment is only part of the work. System designers should also choose lights and appliances that meet their needs – and that consume energy within the design criteria of the system. Energy-efficient (or low-energy) lamp and appliance choices will reduce the overall costs of solar and battery equipment.

The type of lights and appliances that can be installed in an off-grid system depend on two factors: the solar energy available each day; and the voltage of the system.

Low energy lights and appliances make the most out of your PV charge

It makes little sense to spend a lot of money on an off-grid PV system without considering efficient appliances – even if they cost a bit more. A common example is the computer: old desktop computers typically consume three or four times the energy of laptop computers. This means that to properly power a desktop computer one needs to purchase three to four times the amount of solar equipment (modules and batteries). Even if a laptop costs twice as much as a desktop computer, the overall cost of powering it off-grid with a PV system will be less than the desktop computer!

In short, spend some time considering lights and appliances so that they are right for your needs. Do not make the mistake of buying expensive solar equipment without considering the end-use appliances, or buying lamps and appliances without considering how they are to be powered.

The right voltage lights and appliances make a big difference

AC light fixtures and appliances cannot be directly connected in low-voltage DC off-grid systems without an inverter (see Chapter 5). Instead, extra-low energy 12V or 24V DC lamps are commonly used in off-grid PV systems. If possible, especially in smaller systems, avoid using AC appliances and instead select DC appliances, such as DC fridges, TVs or sound systems. Even when an off-grid PV system has an inverter, DC lights and appliances save energy, improve system performance and decrease system complexity.

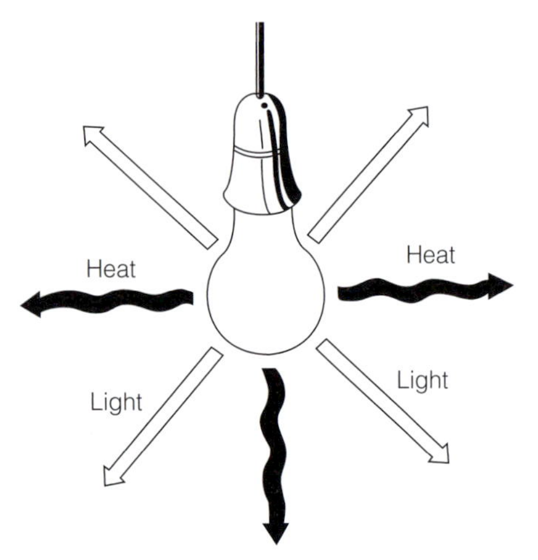

Figure 6.1 *Lamps produce both light and heat energy*

Lighting Principles

Electric lamps convert electric energy into light energy, also called 'visible radiation', which is that part of the spectrum that our eyes can detect. We can see visible radiation, but we cannot see heat radiation or ultraviolet radiation. Efficient lamps give off a maximum of visible light while producing only small amounts of wasted heat. Inefficient lamps such as incandescent globes produce more wasted heat and less visible light.

There are five important lighting principles to consider when choosing and installing lamps:

- the amount of visible light the lamp produces;
- the amount of electric power required by the lamp;
- the voltage required by the lamp, DC or AC;
- the direction in which the light shines;
- the colour temperature of the light.

Figure 6.2 *Important points to consider when choosing lamps*

Lighting efficacy

When a lamp is in use, it produces both heat and light energy (see Figure 6.1). Visible light produced by a lamp is called the 'luminous flux': this is the amount of visible light a source produces measured in lumens (lm). As an example, common kerosene lanterns produce about 100 lumens; a 40-watt incandescent globe lamp produces about 400 lumens; and an 8-watt fluorescent lamp produces 230 lumens. Light falling on a surface is commonly measured in lumens per square metre or lux. A lux-meter is used to measure lighting levels.

Lamps with high efficacies are desirable for PV systems. Electric lamp ratings are always given in watts – this tells you how much power it uses. So, an 8W fluorescent lamp draws 8W of electric power, while a 40W globe lamp draws 40W. 'Efficacy' is a special term used to indicate how much visible light (in lumens) is produced per watt of electric power. Lamps with high efficacies produce more visible light per watt of power than lamps with a low efficacies. For example, the 8W fluorescent lamp described above has an efficacy of 30 lumens per watt (240lm ÷ 8W = 30lm/W). The globe lamp has an efficacy of only about 10 lumens per watt (400lm ÷ 40W = 10lm/W). Compare the lamp types in Table 6.1 to see which have the best efficacy.

Directing light

The right lamp puts light where you want it. Light moves from a bulb in all directions, but it may only be needed in one direction. For example, if you are trying to read a book, then the light should shine on the pages of the book, and not on the ceiling. Light can be directed where it is needed using 'reflectors' (see below). Even a small amount of visible radiation may be useful if it is directed on to the place where it is required. Always look at the lamp fixture you are buying, not just the bulb, because the fixture itself directs the light.

Colour temperature

Lighting applications – and customers – often require light of a certain colour. For example, while it is excellent for working in an office or classroom, the 'cool white' light from fluorescent tubes may not be pleasant in a living room and it may have other undesirable side affects (e.g. attracting insects). In living rooms, a softer, warmer coloured light is often desired. Therefore, when planning solar PV systems select bulbs (and their light colour) to meet the needs of the place where they are used. In some countries lighting levels and types for particular applications (e.g. offices, workshops, street-lighting) are covered by codes and regulations that need to be complied with.

Colour temperature refers to the tone of the light produced. It is measured in degrees Kelvin (K). Incandescent lamps have colour temperatures that range between 2700K and 3000K. Fluorescent lamps have higher colour temperatures (4100K to 6300K) and are often referred to as 'cool' (i.e. 'cool white'). Sources with lower colour temperatures (2700K to 3400K) are referred to as 'warm' (warm white). Normal outside daylight has a 6500K colour temperature. Although fluorescent lamps are commonly rated as 'cool white' or 'warm', 'daylight' tubes are also available, and are often preferable.

Note that all types of lamps (fluorescent, LED, incandescent) come in a variety of colour temperatures and you can choose the colour temperature based on the needs of the application.

Incandescent and Halogen Lamps

Incandescent lamps, or globe lamps (see Figure 6.3), are made with a thin tungsten wire (called a filament) inside a glass globe that contains an inert gas. When electricity passes through the filament it heats up to a very high temperature (i.e. around 3000°C, or 5500°F) and glows brightly, giving off both light and heat.

Incandescent lamps have low efficacies of between 9 and 16 lumens per watt and have a rather short life of between 500 to 1000 hours. Thus, although

Figure 6.3 *Incandescent lamps*

incandescent lamps should be avoided because of their low efficacy, they have some advantages. First, they are much less expensive than fluorescent lamps and they do not require ballast inverters to work (see section below). Also, globe lamps are not damaged by low voltage from the battery (though their lives are shortened when the lamp is operated above its rated voltage).

Halogen lamps are a special type of incandescent lamp with filaments that produce as much as 50 per cent brighter light than regular incandescent lamps (up to 30 lumens per watt) and they last about twice as long. Halogen lamps are often sold with special reflectors that carefully direct light where it is needed (see Figure 6.4).

In general, incandescent and halogen lamps are poor choices for off-grid PV systems. Nevertheless, small PV lighting systems often use inexpensive auto globe lamps in places where light is only needed occasionally or for short periods such as storerooms, bathrooms and hallways.

Fluorescent Lamps and Ballast Inverters

Fluorescent lamps, also called tube lamps or compact fluorescent lamps (CFLs), use current flowing through mercury vapour to produce light radiation. A fluorescent lamp is a glass tube containing mercury vapour and argon gas, with electrodes at either end of the tube. When the lamp is turned on, an electric current flows from the electrodes through the mercury vapour in the tube. The current causes the mercury vapour to give off invisible ultraviolet radiation that, in turn, strikes the inside of the phosphor-coated glass tube, causing the phosphor to glow with a bright white light.

Figure 6.4 *Halogen lamps and fittings*

Depending on their size and type, fluorescent lamps have high efficacies of between 30 to 75 lumens per watt or higher. Tube lamps have long lifetimes of between 2000 and 5000 hours. Fluorescent lamps are a good choice for places where light is required for lengthy periods of time. For example, tube lamps work well in dining and living rooms, classrooms and examination rooms in clinics. A wide range of 12V DC tube lamps is available.

Ballast inverters

DC-type fluorescent lamps require a 'ballast inverter'. Because fluorescent tubes operate at high-voltage alternating current (70–100V AC) and cannot be powered directly by 12V DC, low-voltage DC fluorescent fixtures contain a ballast inverter which converts direct current into alternating current and transforms the battery voltage from low-voltage DC to 70–100V AC. Similarly, ballast inverters raise the frequency of the current and they may contain a special circuit that helps the lamp start (see Figure 6.5). Some ballast inverters are damaged by low battery voltage.

Several types of fluorescent lamps are available:

- The 'batten lamp' is a straight tube fastened to a fixture in which the ballast is contained (Figure 6.5).
- The 'compact' or 'PL-type' fluorescent has a tube bent in a U-shape.

Usually PL-type fluorescent lamps are more efficient than the batten-type fluorescent lamp (see Figure 6.6 and Table 6.1). Although the compact type is

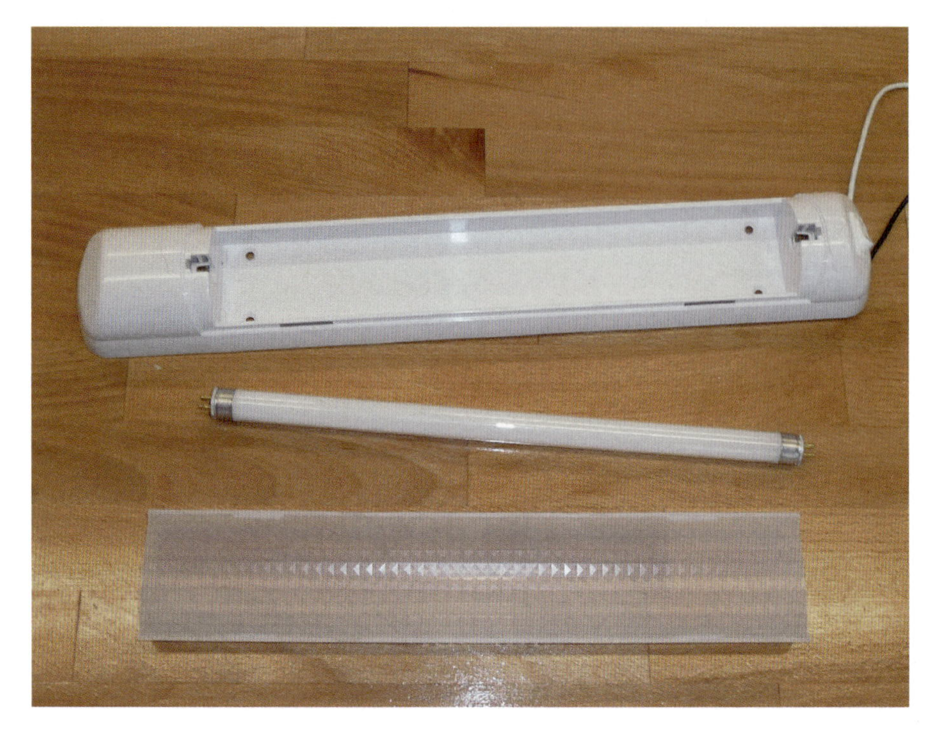

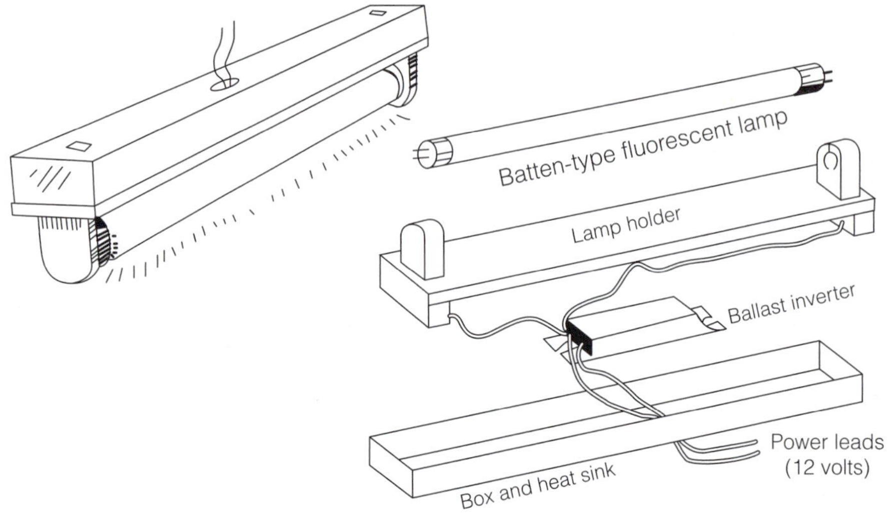

Figure 6.5 *Fluorescent lamp: parts, housing and assembled fitting*

more expensive, it is usually worth the extra cost in terms of better light output. Note also that some PL-type lamps come with replaceable bulbs and others come with integrated bulbs and ballasts (meaning that the entire lamp unit must be thrown away at the end of the bulb's life).

Be careful when disposing of fluorescent bulbs – the mercury in the bulb is poisonous. Better quality fluorescent bulbs will have less mercury content. This is an issue that can be discussed with suppliers, especially when large quantities are involved.

Figure 6.6 *'PL-type' fluorescent lamp fittings. Note that in one model (left in top photograph) the bulb can be replaced (the bottom photograph shows how it detaches) – this version is more expensive than the other model (right in top photograph), for which the whole unit needs to be replaced. Care should be taken when selecting the best light fittings for any situation.*

Figure 6.7 *Range of 'PL-type' flourescent lamp fittings from Steca – 5W, 7W and 11W*

Source: Steca

LED Lamps

Light-emitting diodes (LEDs) use solid-state technology to generate light from electric current without a filament or gas. Materials with LED properties were developed in the early 1960s and initially used as indicators in electric circuitry and panels. Since the mid-1990s, rapid advancements in LED technology have led to vastly increased use of LEDs for lighting applications.

Table 6.1 *Performance of typical 12V lamps*

Lamp type	Rated watts (W)	Light output lumens (lm)	Efficacy (lm/W)	Lifetime (Hours)	Light colour (K)
Incandescent Globe	15	135	9	1000	2700–3000K
Incandescent Globe	25	225	9	1000	2700–3000K
Halogen Globe	20	350	18	2000	2700–3000K
Batten-type Fluorescent (with ballast)	6	240	40	5000	4100–6300K
Batten-type Fluorescent (with ballast)	8	340	42	5000	4100–6300K
Batten-type Fluorescent (with ballast)	13	715	55	5000	4100–6300K
PL-type Fluorescent (with ballast)	7	315	45	10,000	4100–6300K
LED Lamp	3	180	30–100	>50,000	Depends on lamp type

Source: Manufacturers' data

Figure 6.8 *LED (light-emitting diode) lamps. The lamp on the left in the bottom photo uses single LEDs; in the lamps on the right, the LEDs are in 'clusters'. A range of wattages is available.*

LEDs combine a number of advantages for off-grid PV sites:

- long life (50,000–100,000 hours);
- durability (there is no filament to burn out or bulb to break);
- extremely high efficacy (30–100lm/W);
- small size and portability;
- ability to be dimmed (fluorescent lamps cannot be dimmed);
- a wide range of colours.

Because light-emitting diodes often do not produce enough light, LED lamps are commonly available in fixtures that contain clusters of many LEDs (see Figure 6.8). However, newer LEDs also are available that have much higher light-producing power. LEDs are available to meet many types of lighting applications and in a wide range of colours.

Unlike other lamps, individual LEDs only emit light in one direction – this means that LED light sources are less useful for ambient lighting tasks and more useful where direct light is needed. The other main disadvantage of LEDs is their cost, but this is coming down rapidly as more manufacturers enter the market.

Reflecting Light to Where it is Needed

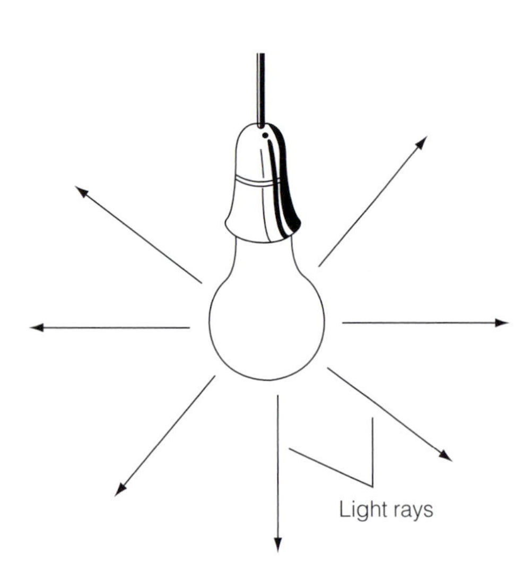

Figure 6.9 *Light travels in all directions from bulb source*

Lamps give off light in all directions. As Figure 6.9 shows, light from a lamp source travels up, down and sideways. If a light fixture is placed in a roof space without reflection, the light rays travelling upwards and sideways are wasted. Reading, sewing, crafts and other work are done below the lamp – those light rays that travel upward do not reach the place where they are needed. In other words, energy is being used to produce light but much of that light is wasted.

Using the principles of reflection it is possible to make use of the light rays that are otherwise lost. A shiny surface causes light rays to bounce off it and return in the opposite direction from which they came. Figure 6.10 shows how a reflector placed around a lamp directs the light rays from the lamp in the desired direction. For example, reflectors found in flashlights direct the relatively small amount of light from the bulb in a beam.

There are two simple ways to make better use of light by reflection in living rooms, work places and classrooms: reflectors and light-coloured paint.

- Reflector fittings are shiny surfaces used to reflect light to the areas where it is needed. They are placed above and on the sides of lamps to reflect otherwise lost light down on to the reading area. Reflectors are made from mirrors, polished stainless-steel sheets or polished aluminium. Reflectors are especially cost-effective in schools and workshops where large areas must be lit. Many lamps come with in-built reflectors.
- Dark surfaces absorb light, while white paint reflects light from the walls and ceilings of a room back into the work area. Rooms freshly painted white are much brighter than rooms with dark or unpainted walls. School and workshop walls should be repainted with white paint every few years to keep the walls bright for easier reading.

Choosing Lamps, Light Fixtures and Mounting Methods

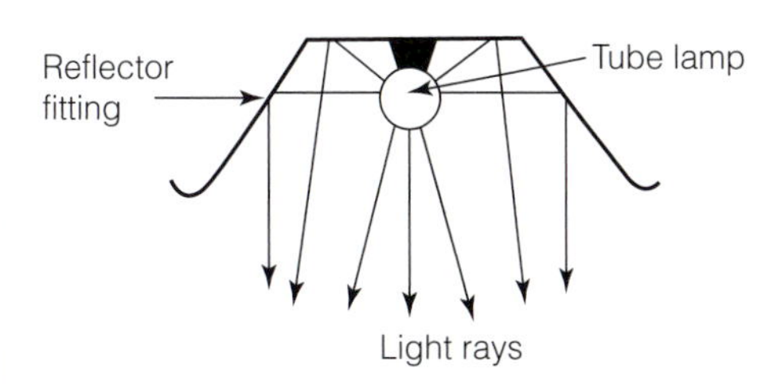

Figure 6.10 *Reflectors direct light to the place where it is needed*

As mentioned above, there is a wide variety of lamps available and the system planner should chose among these carefully. Lamp bulbs and technology are only part of the lighting solution. The type of light fixture used, the method by which the fixture is mounted and the place where the fixture is mounted all contribute to a satisfactory lighting solution. Even the colours of the walls in the room contribute to the overall quality of the light.

The most important factor in the choice of lamps is lamp efficacy. Lamps that use more power require more solar PV to provide enough energy. As mentioned above, fluorescent lamps have efficacies that are four to five times higher than incandescent lamps and they last much longer. LED lamps have even better efficacies and lifetimes. However, there are other factors that influence the lamps chosen, including price, intended use and availability of spare bulbs.

Fluorescent and LED lamps are much more expensive than incandescent lamps. For this reason, off-grid PV system designers who are on tight budgets often include only a few fluorescent or LED fixtures in the house (e.g. in one bedroom, in the sitting room and in the kitchen). Rooms where light is needed for shorter amounts of time may be fitted with incandescent lamps.

The size and choice of lamp also depend on the intended purpose of the light. The amount of light required differs depending on the particular situation for which the light is needed. Whether the light is required for craft work, to light a shop, for security, for study or for social purposes (i.e. ambient lighting) will have a bearing on the size of lamp chosen. The distance from the lamp to the work area is also important because light scatters quickly with distance. The closer a lamp is to the work surface, the less light is lost due to scattering.

Always consider carefully which type of fixture you will use and how you will mount it. Unless they have reflectors, wall-mounted lamp fixtures are good for ambient light, but not task lighting. Fixtures mounted on the ceiling above tables or work surfaces are good for reading or working. Some light fixtures have adjustable turrets that allow light to be pointed in the direction it is needed.

Table 6.2 *Lighting levels in standard situations*

Illuminance	Example
25–50 lux	Bar, dark room during daytime
50 lux	Family living room
80 lux	Hallway/toilet
100–150 lux	Working areas
250–500 lux	Well-lit office or classroom area
750–1000 lux	Workshop, supermarket, operating theatre
10,000–25,000 lux	Full daylight (not direct sun)

Try to choose lamps for which spare bulbs and replacements are available locally. Also, make sure the bulb is the right type for your area (e.g. two-pin baton or Edison screw type).

When choosing lamps for a particular room, the most important factor is the size of the room. Larger rooms, and rooms with high ceilings, require larger lamps. Table 6.2 provides standard lighting levels for various situations that can be measured with a lux-meter. There are often standards for lighting levels in classrooms, offices and clinics. In homes and recreational places, people prefer different lighting levels, colours and types of lighting.

Common Lighting Applications

The examples below provide some guidance on selecting lights for different lighting applications (see also Figure 6.11).

- Health clinic examination rooms require a very strong light so that patients can be observed (>1000 lux). Even small examination rooms require, at a minimum, 13W fluorescent lights. If operations are performed, an even brighter light is required.
- Craft work such as sewing, electrical soldering or beadwork requires more light than ambient lighting (>1000 lux). If light is to be used for such purposes, the system should be designed with directional lamps, such as LEDs, that are adequately sized for the job.
- The large areas of classrooms must be provided with enough light for reading (>400 lux). Experience indicates that students can read well in a typical 8 x 10m (25 x 30 feet) classroom under two 15W fluorescent lamps fitted with stainless-steel reflectors (see Figure 6.11).
- Sitting rooms: owners of off-grid PV homes are often satisfied with lamps placed on low ceilings above a table. This allows those working at the table to read and those sitting in the far corners of the room to see well. One 8W fluorescent tube gives off enough light for a room of up to 4 x 5m (13 x 16 feet). Larger rooms can be lit with two or more 8W lamps with separate switches.

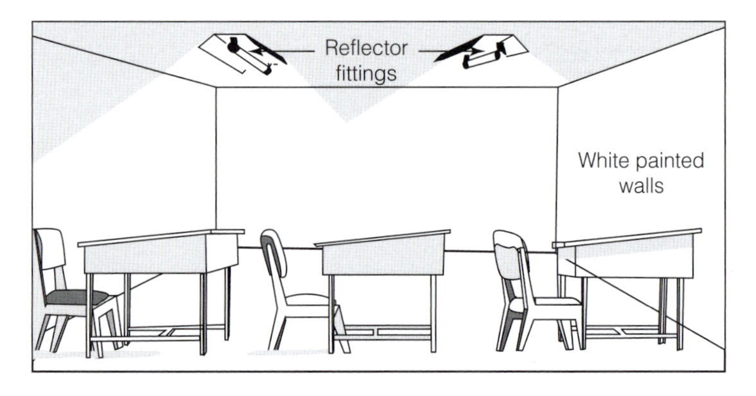

Figure 6.11 *Lighting in a classroom*

- Security lights should be enclosed in weatherproof casings to prevent damage by water or insects. When installing security lights outside homes or schools you should be certain that there is enough solar charge to power them if they are to be used all night. 12V DC movement detectors are available. LED lamps are good for security lamps.
- Street lighting systems installed off-grid are ordinarily designed as free-standing units with their own solar modules, batteries and control systems.
- There are a variety of solar-powered portable lanterns (see Figure 6.12) available on the market. They are useful where a full solar PV system is not required or where the lighting need is temporary (e.g. refugee camps, camping). Some lanterns also provide outputs for powering radios or cell-phones. Be careful when selecting portable solar lanterns! There is a wide selection of types and many are designed for occasional use, not daily use!

Figure 6.12 *Solar lanterns: many portable solar lanterns are on the market. Some are designed for camping holidays, while others are designed for constant use in off-grid areas*

Source: Deutsche Gesellschaft für Technische Zusammenarbeit, www.gtz.de

Appliances

Most off-grid PV systems supply energy for appliances as well as lights. Commonly used appliances in PV systems include televisions, DVD players, radios, cell-phone chargers and communication devices, music systems, sewing machines, fans, small tools, refrigerators, office equipment, medical equipment and laptops.

When designing your PV system, it is useful to make a list of all of the appliances that will be used immediately, as well as those that will be purchased in the future. Try to find out the voltage ratings and the power consumption of all of the devices you plan to use. This allows proper sizing of the system and it also allows a decision to be made about system voltage (or whether an inverter is necessary).

Selecting System Voltage: 12V DC or 110/230V AC?

Appliances are available that operate:

- at 12 volt direct current (or 24V DC if the system is wired at 24 volts); or
- through an inverter at 110 or 230 volts alternating current.

When to use DC

Many 12V DC appliances are readily available from solar equipment suppliers. Even super-efficient DC colour televisions are available. If you have a chance, try to select 12V DC appliances for small off-grid PV systems. There are a number of reasons for this:

- DC appliances are often more efficient than AC appliances, especially if they are made for off-grid purposes.
- It is more efficient to power DC appliances directly from the battery than to use an inverter. Inverters are usually less than 80 per cent efficient and they are even less efficient with loads that use motors (e.g. refrigerators). Running a 200W inverter for a colour TV over the course of a night can consume an extra 150Wh (the entire daily output of a 40Wp module in sunny weather).
- If the inverter breaks down or fails, then so will the AC appliances.
- DC-only PV systems are easier to manage – and well-designed smaller systems are to a large degree self-managing. They are less complex, less can go wrong and they are less likely to be over-discharged because of the charge control low-voltage disconnect.
- Because of inverter losses, it is often better to use DC power to operate small loads that must be left on continuously. Examples of such devices include telephones, doorbells, garage-door openers, motion-sensing devices and alarms.

When to use AC (and an inverter)

Many larger off-grid PV systems (i.e. above 200W) will find that having an inverter is useful. As long as the PV charge is properly sized (and the right inverter type is chosen) it may be a good idea to include an inverter in the system. Choose an inverter when:

- You already have an efficient AC appliance (or appliances). If, for example, you have a music system or a laptop that operates on 230V AC, there is no need to throw it away and look for a 12V DC music system. Use what you have through a properly chosen inverter.
- There is no 12V DC appliance available. In isolated countries or regions, it may not be easy to find 12V DC televisions, laptops or radios. Run what is available through an inverter.
- There is a 110 or 230V AC circuit already wired in the building where the PV system is to be installed (which had been, for example, previously run by a generator). Again, as long as the appliances are carefully selected and energy use is managed, an inverter may be right for this type of system.

Choosing Lights and Appliances

In order that they do not drain the battery, appliances should be energy-efficient and low-energy. When choosing appliances, check the labels to find the unit with the lowest power consumption. Old inefficient appliances (e.g. desktop computers, TVs) should be avoided.

Do not use electric appliances with heating elements such as toasters, microwaves, cookers, hair-driers or electric irons. Only very large solar PV systems (with very large battery banks) can power such systems. If you are not sure if an appliance can be powered by a solar PV system, get advice from a qualified technician.

The list below provides advice about some appliances commonly used in off-grid PV systems. Consult an expert if you need more information.

- Music systems: the watts-per-channel rating on music systems is a maximum rating and several times higher than normal consumption. For example, a 150W per channel rated music system is more likely to consume 50W or less when it is operating (or in mute mode).
- Laptops and desktop computers: laptops use about a quarter of the energy of desktop computers. Avoid desktop computers at all costs! Save even more energy by powering laptops directly from DC-DC converters (see Chapter 12) instead of through inverters.
- Televisions: if possible, shop around for an efficient DC unit. Power consumption in televisions varies greatly depending on the type. Low-cost efficient colour televisions are increasingly available.
- Refrigerators: for off-grid PV systems, DC refrigerators are usually a better choice than AC units. Even though DC fridges cost two times more than standard fridges, they consume less than a quarter of the energy. Use compressor-based 12V DC electric fridges, not absorption refrigerators. Sometimes gas or kerosene fridges (not electric) are more economical in off-grid locations. For clinics that use refrigerators for medical purposes or the cold chain (i.e. keeping vaccines cold), it often makes sense to buy separate and complete systems for the refrigerator (so they have their own modules, batteries, etc.). The World Health Organization provides guidelines for medical solar refrigeration. These need to be complied with.
- Cell-phones: try to use 12V DC chargers provided by the manufacturer. Many 12V DC chargers commonly available 'on the street' are inefficient and do not properly charge your phone.
- Office equipment: be careful when purchasing printers, fax machines, copy machines or other office equipment. Make sure they work with your inverter (if you have one) and that they consume power within your system limitations.
- Medical lab equipment: try to consult the manufacturers of the equipment selected (centrifuge, scales, separators, etc). If powering a PV system in a clinic, make sure that thermal equipment such as sterilizers or water baths will not run from the PV system (unless you have planned for it in system design). Liquefied petroleum gas (LPG) is better and cheaper to power off-grid sterilizers (see Chapter 11 for an example clinic lab system).
- Telecommunications equipment: high frequency radios, wi-fi, VSAT (very small aperture terminal) and other remote communications equipment are often designed to be powered by off-grid solar. Consult your equipment suppliers and make sure they know what type of PV equipment you are using.
- Pumps: get the right size and type of pump, and run it from a separate array if possible. There are many sizes of pumps available and they are designed for different uses. For example, high flow/low head applications use different pumps than low flow/high head applications. Consult an expert

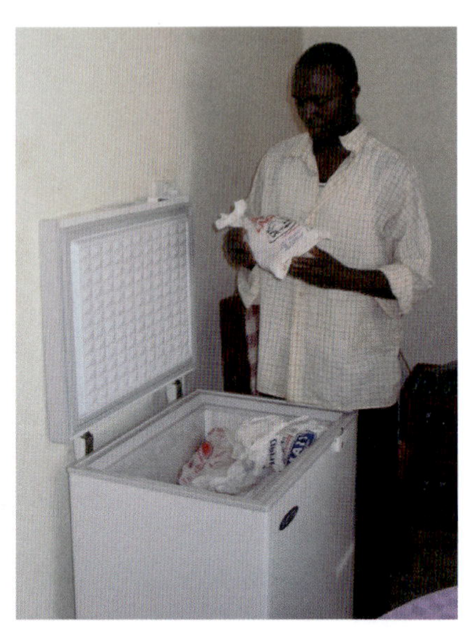

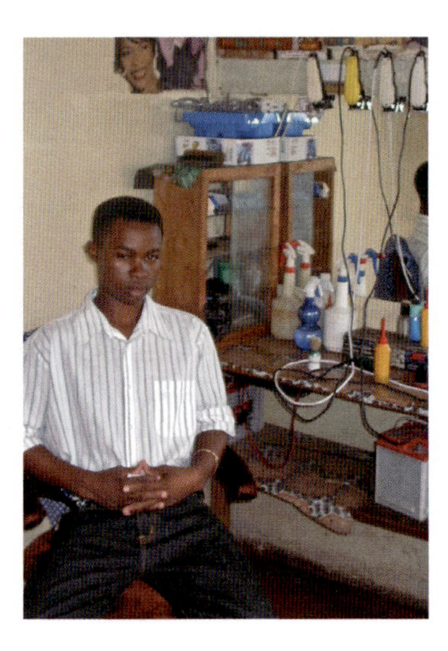

Figure 6.13 *Typical applications often found in small PV systems: 12 volt TV sets; medical and veterinary refrigeration; mobile phone chargers; personal computers (laptops consume a lot less energy than desktops); and inkjet printers (use less energy than laser printers)*

Source: Mark Hankins

Table 6.3 *Approximate power and energy requirements for common off-grid appliances*

Appliance	Typical daily usage time	Power rating (W)	Typical daily energy use (Wh)	Notes
Sewing machine	2 hours	80	50	Motor is engaged only 25% of time
14in colour television	2 hours	80	160	More efficient versions available!
14in black-and-white television	2 hours	24	48	
Radio	3 hours	3–30	9–90	Power draw depends on volume setting
Music system	2 hours	10–40	20–80	Power draw depends on volume setting
Electric iron	30 minutes	300	150	Not recommended for PV systems
Soldering iron	10 minutes	200	45	
Electric drill	5 minutes	150	30	
Computer and monitor	2 hours	80–150	160–300	
Laptop computer	2 hours	25–40	50–80	
Fan	continuous	60	1440	
Water pump	3 hours	450	1000	
DC Refrigerator	continuous	100–150	300–450	Actual energy use depends on ambient temperature. Refer to datasheets

when planning water-pumping. See Chapter 12 for more information on solar-pumping.

- Electric drills: use drills with built-in rechargeable battery power-packs. This will help avoid draining your batteries and make the system portable. If necessary, purchase an extra battery pack so you can exchange them when one is drained.

Appliance Energy Consumption

The best way to find out how much an appliance actually consumes is to measure the DC current being supplied by the battery to the appliance – this, multiplied by the battery voltage, will give the actual number of watts the appliance is consuming (see Table 6.3). The Kill-a-watt energy meter (see Chapter 12) is a useful device for measuring appliance outputs and conducting energy audits.

7
Wiring and Fittings

This chapter describes extra-low-voltage DC cables and fittings for small solar electric systems. Topics covered include: choosing cable size and type, choosing fittings (switches, fuses, connector strips, etc.) and methods of earthing (grounding) the system. Tips on connecting cables are given. The theory of voltage drop is also explained. Two methods of choosing proper cable size are presented with examples.

Remember, this chapter is a general guide for wiring practice in small systems. National codes and regulations, as well as the instructions in equipment manuals, need to be referred to in all cases. The scope of this book mainly covers systems below 500Wp. Wiring, selecting fuses and earthing for small DC-only systems is relatively straightforward. However, for more complex, larger systems, which may involve AC circuits, there is a need for increasingly specialized electrical skills – get advice and assistance from an expert if you do not have skills yourself.

Previous chapters discuss the parts of solar electric systems. Harvested electricity is distributed between these parts using electric cables and fittings. To make efficient use of the energy collected by the modules and stored in batteries, you must choose the right size and type of cables and fittings. Wiring procedures are broadly similar in extra-low-voltage DC systems and in higher-voltage AC systems, but there are important differences, as explained below.

Houses that have previously been wired for AC mains or generator power can be adapted to use solar electric power, while houses wired for DC solar PV can also later be adapted for grid power, depending on how the wiring is done. However, when changing from generator or mains wiring, cables may need to be replaced because solar PV systems often use larger diameter cables. Also, DC and AC fixtures, fuses and switches are quite different – in general, they are not interchangeable.

Box 7.1 Advice for wiring

Selection of correct cables, fuses and accessories for electrical distribution systems can only be discussed in general terms in this book. National wiring codes and regulations need to be complied with, as well as equipment manuals. A good solar system supplier can also provide assistance and advice. Systems should be designed and installed by appropriately qualified electricians. See Chapter 12 for information about common Codes of Practice and standards used in solar PV systems.

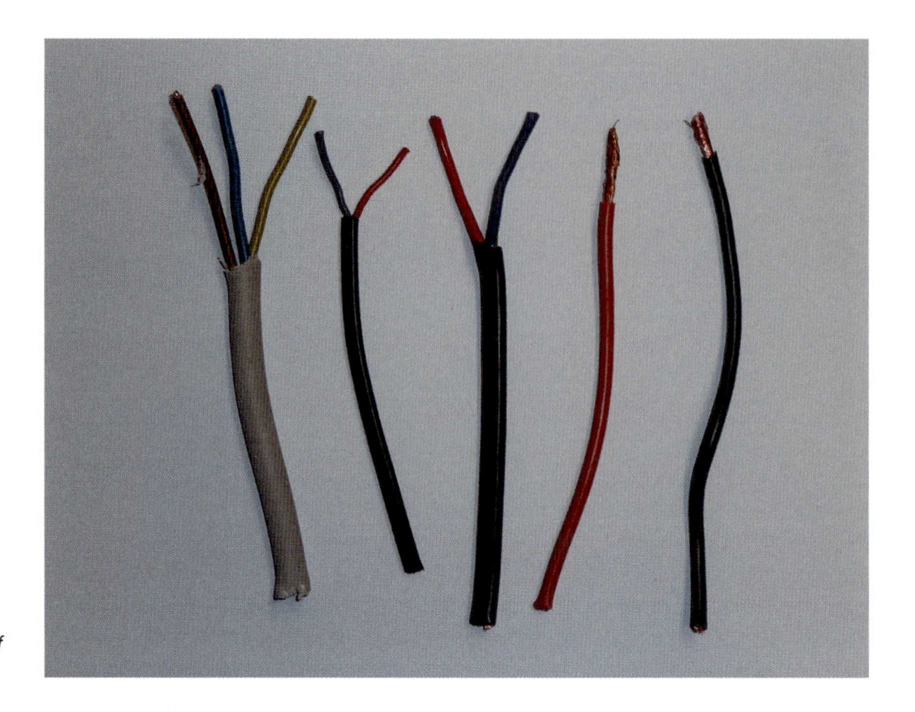

Figure 7.1 *Common types of wiring cable*

Wiring Cable

Make sure to choose the proper cable size and type when planning your system (note that the terms 'wire' and 'cable' are used interchangeably in this chapter). Copper cable is available in various sizes: $1.0mm^2$, $1.5mm^2$, $2.5mm^2$, $4.0mm^2$, $6.0mm^2$, and $10.0mm^2$. (See Figure 7.1). This size refers to the cross-sectional area of the wire; in the US and Canada, the American Wire Gauge, AWG, is used to indicate cable sizes.) Table 7.1 lists common international and North American sizes of cables and their resistance values (see below).

While AC systems normally distribute power using smaller sized cables (i.e. 1.0 or $1.5mm^2$), 12V DC systems ordinarily use cable of at least $2.5mm^2$ or larger. Flexible/stranded two-core cable of $2.5mm^2$ size is ideal for many 12V

Table 7.1a *Copper cable size and resistance (metric sizes)*

Cable Size	Resistance in ohms per metre (Ω/m)
$2.5mm^2$	0.0074
$4.0mm^2$	0.0046
$6.0mm^2$	0.0031
$10.0mm^2$	0.0018
$16.0mm^2$	0.0012
$25.0mm^2$	0.00073
$35.0mm^2$	0.00049

Table 7.1b *Copper cable size and resistance (American sizes)*

American Wire Gauge (AWG)	Size in mm^2	Resistance in ohms per foot (Ω/ft)
14	($2mm^2$)	0.002525
12	($3.31mm^2$)	0.001588
10	($6.68mm^2$)	0.000999
8	($8.37mm^2$)	0.000628
6	($13.3mm^2$)	0.000395
4	($21.15mm^2$)	0.000249
2	($33.62mm^2$)	0.000157
1	($42.41mm^2$)	0.000127
0	($53.5mm^2$)	0.000099

Note: These measurements are for a one-way run, meaning they must be doubled when making resistance calculations.

distribution circuits, and larger sizes are also available. Stranded wires, being considerably more flexible, are easier to work with and fix to walls, and are used a lot in smaller solar home systems. Choosing the right cable is important from a safety and ease-of-work point of view. If in doubt get advice.

Three-core cables – wires that have a separate ground or 'earth' wire – are used in larger DC and AC systems (i.e. for circuits that serve DC and AC equipment receptacles respectively). Again, note that use of earth/ground cables requires qualifications and knowledge of local codes and standards! See the section in this chapter on earthing.

Be careful when using colour codes to indicate DC polarities in cables. Different countries use different colour codes for polarities. The two most common colour combinations used are: brown (+) and grey or blue (–); or red (+) and black (–). Earth wires are mainly green/yellow. When selecting cable colours always keep to local standards. In the case of appliance cables (which may not match local standards), check the instructions for information. Table 7.2 below shows the current IEE (Institute of Electrical Engineers) DC cable marking and colour codes.

Table 7.2 *IEE DC cable marking and colour codes*

Function	Marking	Colour	Note
Two-wire unearthed DC power circuit			
Positive of two wire	L+	Brown	Previously red, still used in places
Negative of two wire	L–	Grey	Previously black, still used in places
Two-wire earthed DC power circuit			
Positive (of negative earthed) circuit	L+	Brown	Previously red, still used in places
Negative (of negative earthed) circuit	M	Blue	Previously black, still used in places
Positive (of positive earthed) circuit (rare)	M	Blue	
Negative (of positive earthed) circuit (rare)	L–	Grey	

In reality, installers use cables that are locally available. If these are not the right colour, mark the colour code or label cables (+ and –) using insulating tape. Remember, the point of colour codes is to make sure that anyone modifying or repairing the system afterwards knows which cables are positive and negative. AC cables are also colour-coded, but that subject is beyond the scope of this book.

Just because cables are carrying 12V DC or 24V DC does not mean that installation standards can be lower than they are in 230/110V AC house mains wiring! A short circuit in a 12V DC system can cause a fire as easily as in an AC circuit. The larger the system the larger the risk. Large DC distribution systems are often installed in hospitals and schools. These circuits need to be installed and tested as specified in electrical codes using the correct instruments. See Chapter 9 for a discussion of this subject.

Some important points to remember:

- Cables laid where they will be exposed to the sun should either be specially selected sunlight-resistant cable (IEC 60811) or they should be encased in the right type of sun-resistant conduit, which is a special type of plastic pipe used for enclosing electric wire.
- When DC cables pass underground or up outside walls, they should be run through a conduit (or they should be armoured).
- Always refer to and comply with local codes.

Switches, Sockets and Fuses

Switches

Switches are used to turn appliances and other loads on and off. They also serve the important purpose of disconnecting modules, batteries and loads during servicing and emergencies. Always select the right type and size of switch for the purpose.

Switches and disconnects need to be properly rated for the circuit in which they are being installed – in terms of current and voltage. A switch or disconnect in a 12V DC circuit needs to be rated for 12V DC and the maximum current expected in that circuit, while a switch or disconnect in a 230V AC circuit needs to be rated for 230V AC and the maximum current expected in that circuit. Many switches and disconnects are rated for both DC and AC current/voltage, though the values for AC and DC may be different. When 230V AC switches must be used to turn lights or small appliances on and off (e.g. because suitable 12V DC switches are not available, which is often the case) always make sure that their nominal current rating is twice the maximum expected DC current.

Only use the proper DC-type switches of the correct voltage and current rating on main switches that control high current DC appliances, PV array or battery circuits. Improperly used AC switches may burn up or arc, and may cause dangerous short circuits or fires! (See Figure 7.2.)

Sockets

Sockets (or power outlets) are receptacles into which the plug is inserted to access power for appliances. In DC circuits 'DC sockets' must be used. Mixing

Figure 7.2 *Common AC switches for lighting circuits*

up AC and DC sockets and plugs can be extremely dangerous. DC sockets are a different shape to AC sockets to prevent AC appliances from being plugged into them (see Figure 7.3). Firstly, there is the possibility that the wrong appliances might get plugged in, damaging the appliance. Secondly, AC sockets are designed for low currents and may not be able to handle the high current of DC circuits. Also, it is extremely dangerous if a 12V DC appliance which has been fitted with a 110/230V AC plug is plugged into an AC socket.

Fuses and miniature circuit breakers (MCBs)

Fuses are devices placed in circuits to prevent accidental damage to appliances, modules and charge-controller circuitry from high current normally associated with short circuits. The very high current that batteries will deliver under short-circuit conditions can cause fires, extensive damage or even explosions! Ideally,

Box 7.2 Labelling sockets

Use DC sockets and plugs of the correct current/voltage rating in DC circuits, and AC sockets and plugs of the correct current/voltage rating in AC circuits. Do not mix them up! To avoid accidents, label sockets in 12V DC systems: '12 VOLT DC APPLIANCES ONLY!'

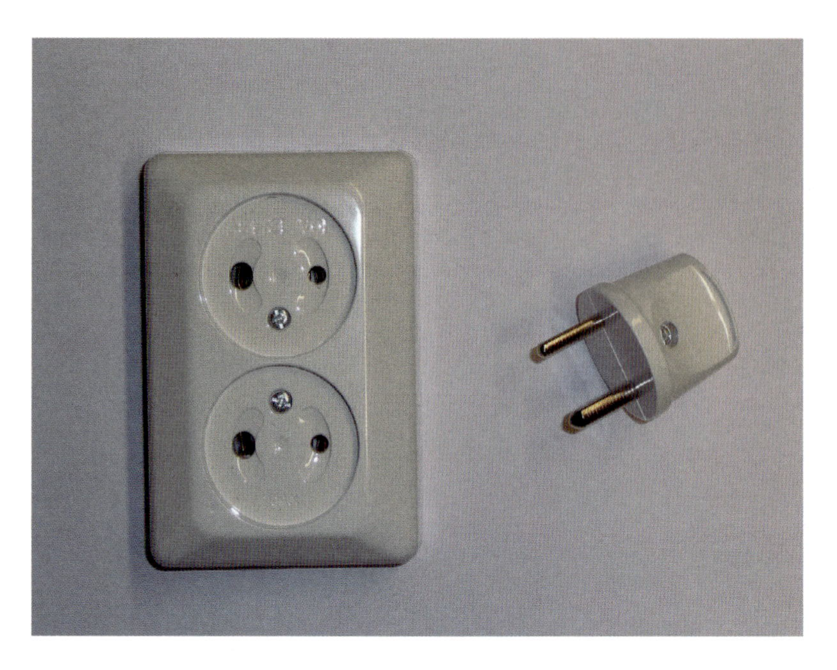

Figure 7.3 *DC sockets and plug*

in a system, there should be a fuse on each of the battery, solar array and load circuits.

When a short circuit occurs, or there is an overload, the fuse 'blows' (i.e. a strip of wire inside melts) and opens the circuit so that current cannot flow. Once a fuse has blown, the cause of the high current should always be investigated and repaired before replacing the fuse with a new one of the same rating.

Miniature circuit breakers (MCBs) are small switches that automatically break the circuit when there is a short circuit or overload. Unlike fuses, they can be switched back on once the wiring problem has been corrected.

DC-rated fuses and circuit breakers should be used in DC circuits, and AC-rated fuses and circuit breakers should be used in AC circuits. They also need to be correctly rated for the circuit voltage.

As a minimum safety precaution, all small systems (less than 100Wp) require at least one fuse: the main battery fuse. Larger systems should have a fuse to protect each major circuit, the battery and the module/array. If there are loads that need to be protected independently, then fuses should be included in the circuit of that load.

Some charge controllers contain in-built electronic load and circuit protection. Look for these charge controllers that have circuitry to protect loads and PV arrays. Such charge controllers not only avoid the problem of including multiple fuses, they also avoid the common (and very dangerous!) practice of consumers replacing blown fuses with the wrong-sized fuse wire.

In all cases, when planning fuse protection, choose the main battery fuse first and follow these suggestions:

- The fuse should be DC rated.
- It should be on the positive cable(s) from the battery, as near as possible to the battery's positive terminal in unearthed and in negative earthed-systems, which most systems are.

- Its rating in amperes (A) should be less than the thermal rating (current rating) of the battery cables. A 30A fuse protecting a cable designed to take 20A means that if 29A flows in the cable the fuse will not blow – but the cable, which is designed to take a maximum of 20A, will overheat and become a fire hazard. However, a 15A fuse would provide full protection.
- Its 'breaking capacity' (in kA) should be greater than the battery short-circuit current. This means that the fuse needs to be able to blow (i.e. not arc) if there is a short circuit – short-circuit currents can be very high.

Other fuses (often located in the charge controller) are important but do not protect against battery short circuit from faults in cables between the battery and the charge controller. When placing fuses on inverter circuits refer to the inverter manual, which should specify fuse size and type (as well as recommended cable size from battery to inverter).

Sizing fuses

Sizing fuses is not straightforward but national electrical codes and inverter/battery/charge-controller manuals will give specifications. The instructions below are only for: battery fuses in single battery systems below 100Wp; and fuses in 12V DC distribution circuits. However, they will give a introduction to the process. To calculate the required fuse size for each major 12V DC circuit, follow the steps below and use Worksheet 5 (found in the Worksheets section, page 225, and available as a spreadsheet at www.earthscan.co.uk/expert). Note that AC system fusing and breaker design for inverter circuits are beyond the scope of this book. Consult an electrician when selecting AC fuses.

1. List the circuits that need to be protected in Column A in Worksheet 5 (Step 3: Sizing Fuses and Circuit Breakers).
2. Divide the power by system voltage to get current in amps. This is the maximum rated current in the circuit (Column C).
3. Increase this figure by 20 per cent of its value (i.e. multiply this figure by 1.2, see Column D). This is the size of the fuse required.

Remember fuses also protect cables from heating up (a fire hazard), so the 'current rating' or 'cable thermal rating' of all cables protected by this fuse need to be greater than the current rating of the fuse.

If there is only one fuse in the system, then cables serving the lights and the appliances and connected to the modules are also protected by it – but only if the current rating or cable thermal rating of all cables protected by this fuse are greater than the current rating of the fuse. It is often necessary to place a fuse in each major load circuit.

Selecting fuses

Fuses are rated in amps. They are sized to 'blow' very quickly when the current is about 20 per cent greater than the maximum expected current in the circuit. If, for example, there is a short circuit in one of the appliances, the circuit draws

Box 7.3 Battery fuse sizing example

A 12V system circuit includes a 60W television and three 10W lamps:

60W + 10W + 10W + 10W = 90W

Divide the power by the system voltage (12V):

90W/12V = 7.5A

This is the maximum current coming from the battery to the charge controller. Increase this figure by 20 per cent:

7.5A × 1.2 = 9A

In the above circuit example, a 9-amp fuse should be used. Since this size does not exist, select the next largest size, in this case a 10-amp fuse. The current rating of the cable needs to be greater than 10 amps.

much more than the rated current (i.e. more than 20 per cent higher), so the fuse rapidly heats up, 'blows' very quickly and opens the circuit.

PV suppliers and electronic equipment shops stock fuses of various sizes, ranging from 0.25 amps to 30 amps or larger (see Figure 7.4). Make sure you select the right fuse and keep enough spares to replace the blown fuses. Ask your PV dealer to provide you with the right type of fuse.

For extra-low-voltage DC circuits of 5 amps and above, cartridge-type fuses are commonly used. Below 5 amps, glass 'automotive' type fuses are often used. Avoid 'wire-type' fuses as they are easily bypassed and not appropriate for solar PV systems.

Automotive fuses are designed for use with SLI or sealed 'leisure' types of batteries. Although not specifically made for solar PV systems, they are integrated into some types of charge controllers for small SHS (i.e. less than 100Wp) and even small caravan lighting systems. Use of automotive-type fuses should be limited to systems that contain only one 12V battery. If more than one battery is to be used then a 'proper' cartridge-type fuse is needed.

Be careful when choosing 'AC-type' fuses for DC applications. They may also be rated for DC, but at a lower current. The DC rating is commonly given in catalogues and datasheets, but is often not on the fuse itself. DC-rated MCBs are also available.

Making Connections

Wires in solar electric systems should be connected securely, safely and carefully. The extra-low DC voltage of solar PV systems is not an excuse to take shortcuts with regards to safety. More system problems are caused by bad connections than by failures of the equipment itself. The following tips are given to help make sure that the initial connections last a long time. But remember, good wiring is a skill, and needs to be learned.

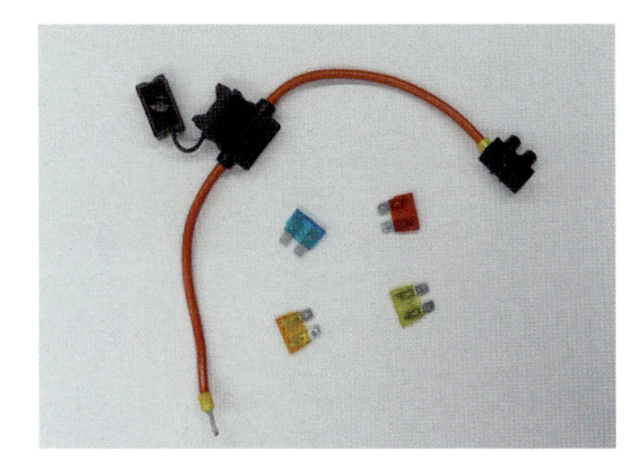

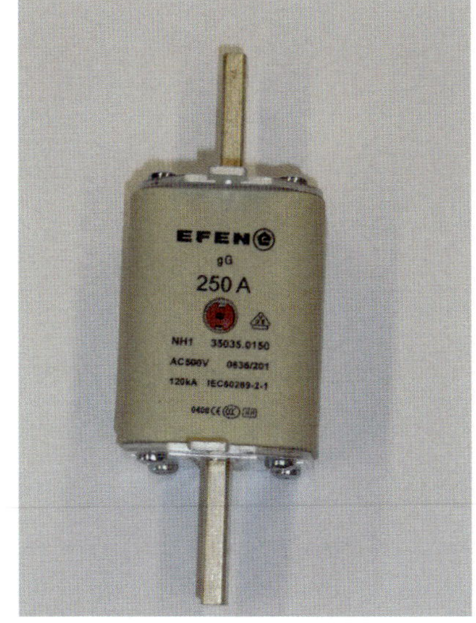

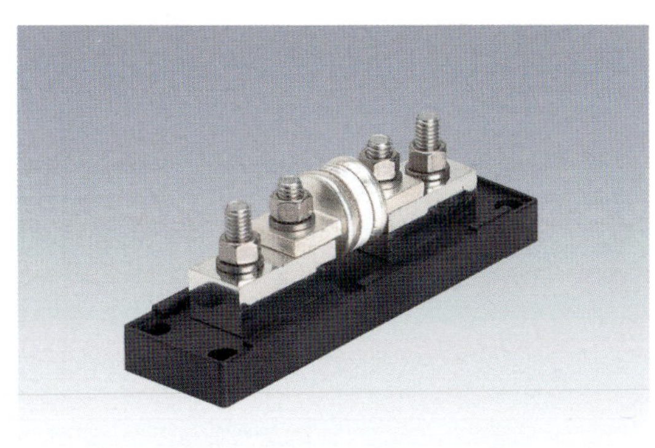

Figure 7.4 *Battery DC-rated fuses: the blade type fuses (top) are only suitable for a single 12V battery of up to about 100Ah capacity; larger battery banks need the types of fuses shown in the bottom two pictures*

Use connector strips and/or junction boxes

Connector strips are insulated screw-down wire clamps used to connect wires together (see Figure 7.5). Connector strips are available in various sizes to fit 2.5mm^2 or larger wire. Junction boxes are plastic or metal containers inside which electrical connections are made. They are normally mounted inside walls. Buy the right size junction box for your installation and make connections properly. Never connect wires by twisting and always enclose connector strips in boxes.

Prepare wire ends carefully

Strip 0.5 to 1cm (about 0.25 to 0.5 inches) of insulation from the end of the wire. Make sure the wire is clean. Then, before fixing the wire to a terminal or connector, twist the end. If possible use 'shoelace crimp terminals' (see Figure 7.6).

Use weather-proof boxes and conduits when connecting wires outdoors

If connected outdoors, wires should be enclosed in junction boxes. Make sure there is extra wire for entry and exit from junction boxes.

Figure 7.5 *Connector strips and junction boxes*

Figure 7.6 *Shoelace crimp terminals*

Do not twist wires around terminal connections in the battery, module or charge controller

Use a crimping tool (see page 149), if one is available, to fix ring-type or spade-type ends to the wire. These are less likely to be pulled off or to be affected by corrosion. If the installation site is near the ocean, solder terminal connections so that they do not corrode.

Inspect all connections carefully after installing

Make sure no wires are loose. Check for places where bare wire might overlap and cause a short circuit.

Be neat in wiring

Neat wiring not only looks better, it is easier to service and less likely to get tangled, or crossed and shorted. Align wires coming from terminal strips so that they are straight.

Seal all fittings and switches

This should be done with a silicon paste (non-flammable – ask your electrical supplier for advice). This is to stop insects from getting into junction boxes, light-fittings and other items of electrical equipment.

Wire Size, Voltage Drop and Maximum Wire Runs

Solar PV system designers must choose whether to use 12/24V DC circuits, 110/230V AC circuits or some combination of the two. This book is primarily concerned with systems up to 500Wp that use 12/24V DC circuits and the wiring practice of lower voltage DC systems. Larger AC and combination AC/DC systems are also briefly discussed, but require assistance from qualified electricians (see Chapter 8).

In extra-low-voltage DC systems, selecting the correct cable size is extremely important. As with water pipes, the cross-sectional area of a cable determines how much current can pass through it. Larger diameter cables allow more current to flow than cables with small diameters (see Figure 7.7). Because lower voltage systems require a higher current to carry the same power as higher voltage systems, they generally require larger sized cables.

Box 7.4 shows the difference in current flow between wires in a 24W lamp connected to a 12V DC battery and a 24W lamp in a 230V AC grid system.

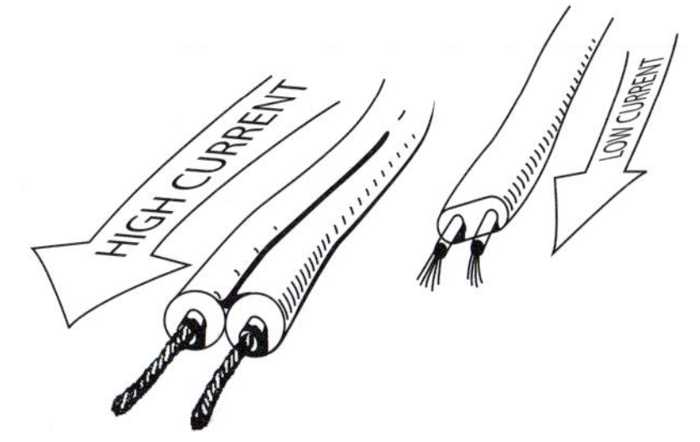

Figure 7.7 *Larger-diameter cables are needed to carry larger currents*

What is Voltage Drop?

Voltage drop is the loss of voltage (and hence power) due to resistance in long cable runs. If the wire's cross-section area is too small for a given current, an unacceptable voltage drop will occur over its length. Resistance in the cable converts electrical energy to heat and causes a consequent voltage drop. When the voltage drop is too large in, for example, the cables from the PV module or array, the battery or battery bank will not be charged properly; in distribution circuits it will affect performance of lamps and appliances, and may damage them. The voltage drop also wastes expensive energy from the PV array and battery.

Voltage drop occurs in all wire runs. However, voltage drops of more than 5 per cent are always unacceptable. Correctly selected cable sizes will avoid unacceptable voltage drops. Review every circuit in an extra-low-voltage system for voltage drop.

Box 7.4 Current flow in extra-low and low voltage systems

The Power Law (see Appendix 2) is used to determine the current flowing in each wire:

Power (watts) = Voltage (volts) × Current (amps)

A 24W globe lamp consumes 24W of power regardless of whether it is designed to work with a 230V AC or 12V DC power supply. However, to produce the same power as higher voltage AC, more current must flow in the lower voltage lamp. Figure 7.8 shows the current draw of 12V DC and 230V AC voltage lamps. Note that the current drawn by the higher voltage lamp, 0.1A, is 20 times less than the extra-low-voltage lamp's current of 2.0A.

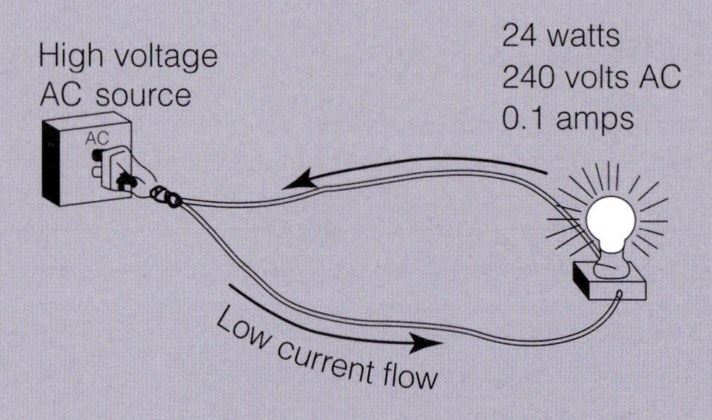

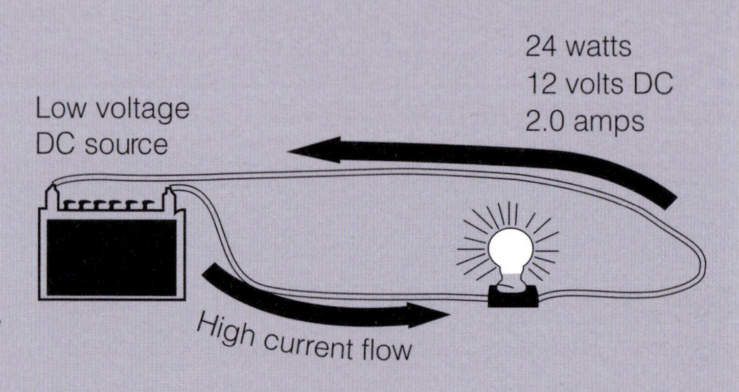

Figure 7.8 *Current flow in 240V AC and 12V DC systems*

The above example should make it clear why lower voltage systems use a much higher current than higher voltage systems, and why they require larger cable sizes.

Suppose, for example, that a 24W lamp is powered from a battery located in a kitchen 50m (165 feet) from the battery (see Figure 7.9). How much voltage drop will there be if a 1.5mm^2 cable carries the power to the lamp? Will the voltage drop affect the performance of the lamp? If so, what is the correct wire size to be used?

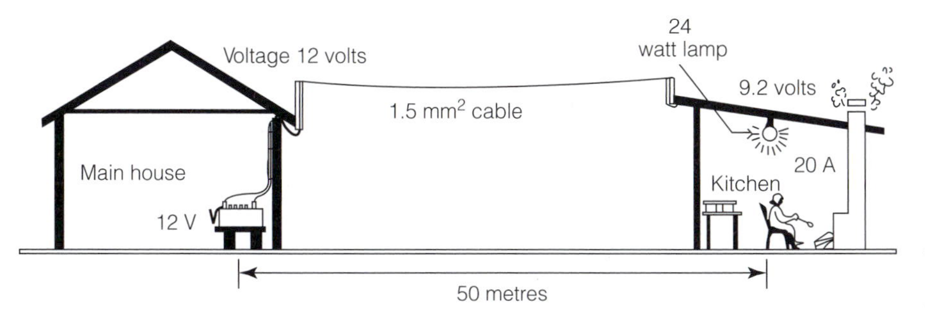

Figure 7.9 *Voltage drop in a 12V DC system: the cable between the buildings is too small*

Box 7.5 When it isn't necessary to calculate voltage drop

In very small systems, all connections can safely be made using 2.5mm² wiring cable if all three of the following conditions are true:

- no wire run is more than 8m (26 feet) long;
- the module is rated at 40Wp or below; and
- no wire carries a current greater than 4A.

Table 7.3 *Suggested maximum permissible cable voltage drops*

Battery to charge controller	<1%
Battery to inverter	<1%
Solar module to charge controller	<3%
Charge controller to loads	<5%
Inverter to loads	<5%

These questions need to be answered before installing the lamp. Whether voltage drop will affect an appliance depends on the appliance (fluorescent lamps may be damaged when run at a voltage below the recommended level). See Table 7.3 for suggested maximum permissible cable voltage drops.

Using Worksheet 5 to Calculate Voltage Drop

Proper wire size for a cable run can be determined in either of two ways: by using maximum wire run tables; or by calculating the voltage drop. Both methods are described in the following sections. For both methods, use Worksheet 5, which is reproduced below in Table 7.4.

Maximum wire run tables

1 List each wire run together with its one way distance. Measure the actual length of required wire run (i.e. bends, extra distance around objects, not just the straight line distance). Write these in Columns A and B in Worksheet 5.

Table 7.4 *Voltage drop calculation table*

Column A	Column B	Column C	Column D	Column E	Column F	Column G
Cable run (list circuits)	**Distance of cable (m)**	**Maximum current (A)**	**K Value of wire (ohms/m)**	**Total resistance (ohms)**	**Voltage drop (V)**	**Voltage drop (%)**

2 Find the maximum current passing through each wire. To do this, add the power in watts of all the lamps and appliances on each wire run, and divide by the voltage. Write this on the worksheet in Column C.

3 Maximum wire run tables (Tables 7.5 and 7.6, see below) give the maximum run of cable (one way) that can be used between modules, batteries and loads in 12 and 24 volt systems under various currents.

Use Tables 7.5 and 7.6 as described below:

1 Firstly, calculate the current in amps that the wire will carry (as entered in Column C of the worksheet). For example, if a 12V system has two 8W lamps at the end of a wire, the current being carried by the wire is 1.3 amps (8W × 2 lamps ÷ 12V = 1.3A).

2 From the table, read the load current in amps that is closest to this figure. If the current is not listed in the tables, read the next higher figure. (For example, in Table 7.5, if the current to be carried is 3.5 amps, read down the 4-amp column).

3 Read across the row to get the maximum length of cable that can be used for a given wire size. In a 12V system, for example, if the current to be carried by the cable is 4 amps, the maximum wire run distance of a 2.5mm^2 cable (with below 5 per cent voltage drop) is 9m (30 feet).

Table 7.5 *12V system maximum wire length in metres (0.6V max. voltage drop, or 5 per cent)*

Wire size (mm^2)	Load current								
	1A	**2A**	**3A**	**4A**	**5A**	**6A**	**8A**	**10A**	**14A**
1.5	22	11	7	6	4	4	3	2	2
2.5	38	19	13	9	8	6	5	4	3
4.0	60	30	20	15	12	10	8	6	4
6.0	88	44	29	22	18	15	11	9	6
10.0	150	75	50	38	30	25	19	15	11

Table 7.6 *24V system maximum wire length in metres (1.2V max. voltage drop, or 5 per cent)*

Wire Size (mm^2)	Load current								
	1A	2A	3A	4A	5A	6A	8A	10A	14A
1.5	44	22	15	11	9	7	6	4	3
2.5	75	38	25	19	15	13	9	8	5
4.0	120	60	40	30	24	20	15	12	9
6.0	176	88	59	44	35	29	22	18	13
10.0	300	150	100	75	60	50	38	30	21

Calculating Voltage Drop Mathematically

Instead of using the above tables, it is possible to calculate the voltage drop on a cable run using simple electrical principles. Ohm's Law states that voltage is equal to the current in amps multiplied by the resistance in ohms (see Appendix 2). This can be used to calculate the voltage drop in a circuit as follows:

Voltage Drop (volts) = Current (I: amps) × Resistance (R: ohms)

To calculate voltage drop, one must first know three values:

- the current flow through the wire in amps (I);
- the distance of the cable run in metres (the length of both the positive and negative runs must be included in this calculation!); and
- the resistance factor, K, of the intended cable in ohms per metre (see Table 7.1).

Multiplying the distance of the wire by the resistance factor gives the total resistance, R.

Box 7.6 Example of voltage drop calculation, table method

A 24W lamp draws 2.0 amps at 12 volts. Reading down the 2.0 amp column in Table 7.5, the maximum distance that a 2.5mm^2 cable carrying 2 amps could support without a 5 per cent voltage drop is 19m (62 feet). The maximum distance a 4mm^2 wire could run without significant losses is 30m (100 feet). A 6mm^2 cable would safely be able to carry 2 amps up to 44m (144 feet).

Note from Table 7.6, that if the lamp was run on 24V DC, the voltage drop is much less, and that even a 1.5mm^2 wire could be used to transmit the current 22m (72 feet) at that voltage. Increasing system voltage decreases voltage drops in long runs.

The voltage drop in each wire is calculated, using Worksheet 5, as follows:

1 Identify each cable run in Column A and note its length in Column B. (Be sure to multiply the one-way distance by two to account for both wires in each cable!)
2 Determine the maximum current that each cable will carry as described above and write this in Column C.
3 Determine the resistance factor, K, of the intended cable using Table 7.1. Note this value in Column D. For example, a 2.5mm^2 copper wire has a K value of 0.0074 ohms per metre.
4 Calculate the total resistance of the cable by multiplying the resistance factor (Column D) by the distance of the wire (Column B):

$$\text{Total Resistance (ohms)} = \text{Resistance Factor, K (ohms/metre)} \times \text{Length of cable (metres)}$$

Write this value in Column E.

5 Using Ohm's Law, calculate the voltage drop through each cable. The voltage drop will be equal to the total resistance (Column E) multiplied by the maximum current carried in the wire (Column C). If the voltage drop is greater than 5 per cent, use a larger diameter cable.

$$\text{Vdrop (Column F)} = \text{maximum current (Column C)} \times \text{total resistance (Column E)}$$

Box 7.7 shows how to use the worksheet to calculate voltage drop in a systematic manner.

Earthing and Lightning Protection

'Earthing' (or 'grounding' for North Americans) refers to a variety of measures taken to avoid shock hazards, to protect against lightning and to ensure that sensitive electrical equipment operates properly. Unless a site is extremely susceptible to lightning strikes, 12V DC solar PV systems below 100Wp require minimal earthing protection. Solar PV systems that include an inverter or generator have increased electrical safety and fault risks, and should follow national AC system regulations and manufacturer installation instructions.

National standards and codes vary considerably with respect to grounding/earthing of PV systems. For example, in some countries (e.g. South Africa) all PV systems are required to be grounded. When installing your system, be sure you understand the need for and principle of earthing/grounding – as well as the regulations in your country and what exactly they require!

As elaborated in the sections below, earthing serves several fundamental protective and functional purposes:

- It ensures that fuses will blow (or circuit breakers will trip) quickly when there is an electrical fault.
- It ensures that all parts of the electrical system that could become 'live' under fault conditions have the same 'electrical potential' with respect to the earth and cannot cause a shock to consumers.

Box 7.7 Example of voltage drop calculation, mathematical method

Suppose electricity is provided for a 24W lamp in a kitchen 50m (165 feet) from the power source using 2.5mm^2 cable. Using Ohm's Law, it is possible to calculate the voltage drop for a 12V DC and 24V DC system (see Table 7.7).

1. For a 12-volt system the current is 2 amps and for a 24-volt system it is 1 amp.
2. The resistance factor of the 2.5mm^2 cable, from Table 7.1, is 0.0074Ω/metre.
3. The total resistance of the wire is calculated by multiplying the resistance factor by the length of the wire. Note that the number is equal for both the 12 and 24V DC system because the wire is the same in each case. Use 100m (i.e. 2 × 50m) for the calculation because it is a two-way run. The total resistance over the wire run is 0.74Ω (0.0074Ω per metre × 100m).
4. Next, as per Ohm's Law, the total resistance is multiplied by the current carried in the wire. This gives the voltage drop.

Table 7.7 *Example of calculation of voltage drop in 12 and 24V DC systems*

Voltage Drop in a 2.5mm^2 cable	12V DC lamp	24V DC lamp
Current (amps)	2.0A	1.0A
Length of Wire (m)	100m	100m
Resistance Factor, K (2.5mm^2 cable)	0.0074Ω/m	0.0074Ω/m
Total Resistance (K × 100m)	0.74Ω	0.74Ω
Voltage Drop (Total resistance × amps)	1.48V	0.74V
Voltage at Lamp	10.52V	23.26V
Per cent Voltage Drop	12.3%	3.1%

For the 12V lamp, the voltage drop is 12.3 per cent. This drop is too high and a larger diameter wire size needs to be used. With a 24V lamp, the voltage drop is only 3.1 per cent. This example demonstrates why 50m (165 foot) cable runs in small solar electric systems are problematic.

- It reduces the risk of damage from lightning strikes.
- It provides functional earthing to enable correct operation of sensitive electronic equipment that must be earthed to work properly.

The sections below discuss each of the above points and the final section provides advice on setting up earthing circuits. See also Chapter 8 for practical advice on earthing.

System earthing and electric shocks

Small 12 or 24V DC solar PV systems without inverters do not normally pose a risk of electric shock. Therefore, provided they have adequate fuse protection, there is little need to worry about earthing unless national regulations specifically require it.

When any component of a solar PV system operates at 110 or 230V AC, the system must be designed to limit the risk of electric shock across the system and to ensure that a ground fault will trigger a circuit disconnection. Ungrounded inverters or appliances can potentially cause shocks when used improperly or in wet or damp conditions.

Note that there is a big difference between small inverters for powering one or two socket outlets and larger inverters (or inverter-chargers) that power entire electric circuits. Larger inverters normally have dedicated terminals for earth connections – and clear instructions about connecting the earth circuit in the manual.

Appliance and accessory earthing

Metalwork associated with the electrical system and appliances can potentially become 'live' under fault conditions. Such metalwork is normally earthed to keep it at the same 'electrical potential' as the general mass of the earth. Earthing ensures that such metalwork never has a voltage above 50V. Ordinarily, this is achieved by connecting all the earth terminals of metal casings and appliances back to a main earthing terminal via an earth wire (see Figure 7.10). The main earthing terminal is connected to the body of the earth via an earth electrode. Note that the earthing of other metalwork in a building (such as water pipes, gas pipes and metal frames) may also be required; this is known as equipotential bonding. The details of how electrical systems are to be earthed (earth-wire sizes, testing requirements, etc.) can be found in national electrical codes and in installation instructions/manuals. Systems which require earthing need to be designed and installed by appropriately qualified persons.

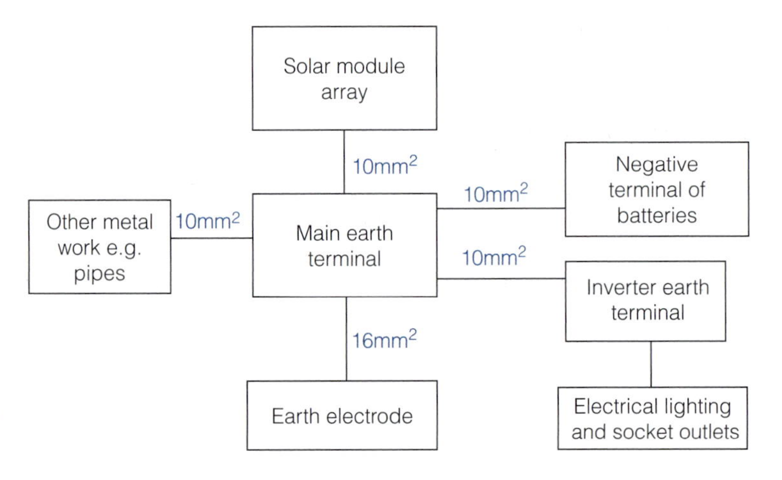

Figure 7.10 *Possible earthing arrangement for an off-grid PV system. Cable diameters are the minimum recommended. Local regulations need to be consulted*

Earthing circuits

Figure 7.11 *Earth rod connection*

In a typical small PV system, all casings of appliances, inverters or arrays and the negative terminal of the battery are connected to a main earth terminal. This, in turn, is connected to an earthed electrode (see Figure 7.11).

Note that:

- Earthing for systems with voltages less than 24V is often not required but national electrical codes need to be consulted.
- Modules and arrays would normally be earthed by connecting the array to the main earth terminal. Where there is a high likelihood of a lightning strike, the array frame may be grounded separately.
- Earthing in systems with inverters and 110/230V AC circuits involves connecting all metal casings and the negative battery terminal to a single terminal, and that terminal to an earth rod. Always read inverter manuals carefully and consult a qualified electrician for advice when planning circuits.

Lightning protection in off-grid systems

In areas prone to electrical storms, protection against lightning is always necessary. Where there is little risk, systems below 100Wp do not usually require lightning protection. Although lightning protection is unlikely to offer protection from a direct hit, proper protection equipment can reduce the effects of voltage, current and magnetic field fluctuations caused by nearby hits. There are two strategies to protect against lightning strikes:

- An external lightning protection system protects the array by attracting the strike and directing it into the ground via protective conductors.
- An internal system, which might be a DC isolator in the charge regulator, reduces the risk of voltage surges damaging other components of the installation.

Functional earthing

Some appliances do not work properly if they are not earthed. This includes some communications and measurement equipment. This is unlikely to be an issue in DC SHS systems. Note that some functional earthing circuits cannot be integrated into central earthing circuits. The IEE colour for functional earth cables is cream.

8

Planning an Off-Grid Solar Electric System

This chapter is a guide to the design steps in planning off-grid PV systems. It combines the information presented in the preceding chapters, enabling you to tailor a solar electric system to your own particular requirements and resources. Worksheets 1 to 5 (printed on pages 221–226 and available online as a spreadsheet at www.earthscan.co.uk/expert) provide a detailed guide through the design steps.

Understanding the Design Process

Make a plan before buying and installing your off-grid PV system. If you don't plan, your solar investment may turn out to be less functional and more expensive than intended!

The five worksheets can be used to design off-grid stand-alone PV systems below 500Wp in size. When designing your system follow the worksheets carefully. However, do not just follow the worksheets mechanically. They are not 'recipes'! They are guides to help designers understand the overall process of putting together PV systems. Practise designing your system a few times using different assumptions. Check the results against each other. Have someone else look at your design. It is good to consult a knowledgeable PV expert or system supplier and to get their input. They will be able to answer your questions and address problems.

Remember:

- No worksheet is perfect. They are not designed to cover all eventualities (e.g. long cloudy periods, misuse by end-users, etc.).
- Worksheets are tools that work like simulation software to predict how a system will function. There is no magic button that one can press to give the 'perfect' system. In the real world solar systems behave according to the weather and the way that end-users treat them.
- These worksheets are not meant to cover hybrid systems. See Chapter 12 for simulation software that can help you design larger systems.

The design steps are as follows (see Figure 8.1):

- Worksheet 1: Calculation of the load energy and selection of system voltage.
- Worksheet 2: Survey of solar resource and selection of PV module(s).
- Worksheet 3: Sizing and selection of battery.
- Worksheet 4: Sizing and selection of inverter and charge controller.
- Worksheet 5: Design of main circuits, sizing and selection, cables, fuses and switches.

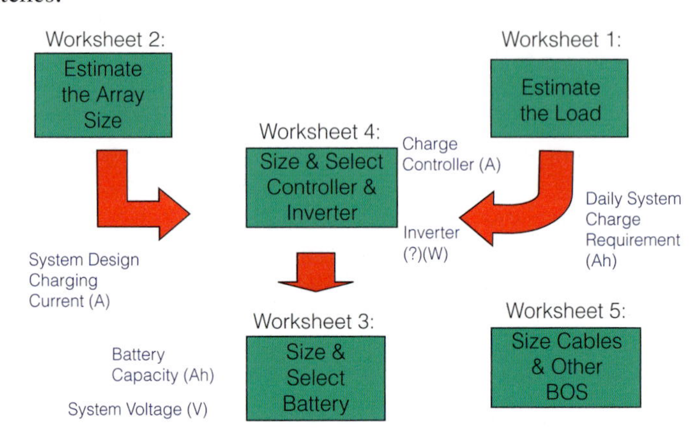

Figure 8.1 *Off-grid PV system sizing process*

In order to show how the worksheets are used, an example is provided at the end of each section. The case study is modelled on an actual residential system installation in Arusha, Tanzania. The entire planning process of the system is presented. It may be useful to review this case before planning your own system. Chapter 11 presents other systems, as well as case studies of hybrid systems and special types of loads (e.g. refrigerators and pumps). In order to follow this chapter most easily, photocopy or print the worksheets and work through them as you read it.

Some Planning Considerations

- The 'design' method presented here is a planning tool that simulates how solar PV systems work. Any simulation is limited because the performance

Box 8.1 Amp-hour v. watt-hour sizing

There are two common methods by which system designers size off-grid PV systems. In this book, the 'amp-hour' method is used (as opposed to the 'watt-hour' method). Both methods give the same answer if done properly. Amp-hour sizing takes into account the real-world behaviour of modules and batteries. It requires the designer to understand the I-V curve of the solar module (as well as effects of temperatures on module outputs).

The amp-hour method is best used on simple small systems where the battery voltage is the same as the load voltage. In systems where MPPT charge controllers are used (see Chapter 5), the amp-hour method is less useful because of differences between battery and module voltages, though one still has to use amp-hours when sizing the batteries. See Chapter 12 for resources on other system-sizing methods and simulation software.

of real-life PV systems varies considerably. The better an owner understands the system, its energy flows and the need for active management, the better the system will work.
- PV equipment requires a relatively large investment. This is a primary reason why people often cannot afford to buy 'perfectly sized' PV systems all at once. If a customer cannot afford to buy everything at once, plan the PV system in phases, starting with fewer appliances and increasing energy use when more modules can be purchased.
- When planning, remember that the electric energy demand will probably grow. For example, in a school, classrooms may be added, or in a home system, a television might be added. Adding more appliances will increase the load and will often require additional solar modules and/or batteries. Plan for this.
- Consider and compare the costs of alternatives carefully. Solar electricity is usually the cost-effective alternative for those far from the grid who require power for lights and small appliances. If a grid transformer is within a few kilometres, it may be cheaper to install the grid (unless the grid is unreliable or very expensive). For some applications, a combination of a diesel generator with solar PV may be practical (see Chapter 11). If possible, investigate the possibilities of other renewable energies, including biogas, solar water-heaters, micro-hydroelectricity and wind-generated electricity.
- Visiting other PV systems is a good way to learn about system design and how systems really perform. If there are other systems in the area, it is advisable to find out what type of problems the users have experienced and to learn from them.
- Before starting, draw a scale diagram of the floorplan where the system is to be installed (or get it from the architect). It will help to estimate the amount of cables and wiring material required, and to think about where best to install the battery, modules, control, sockets, switches and appliances (as discussed in Chapter 9).
- Keep systems as simple as possible. The more complex you make it, the more things there are that can go wrong.
- Shop around for different equipment. Check prices and types of equipment available, and decide carefully which is the most practical (see Chapter 12 for more information on this). The more informed you are about what is available, the better the deal you will get.
- Estimate loads carefully and realistically. Over-estimation of the load will increase the cost of the system significantly. Under-estimating the size of the load may result in continual battery problems.

System Voltage

'System voltage' is the nominal voltage at which the batteries, charge regulator and solar array operate. Also, system appliances often operate at the system voltage. This book is written mainly for those using and installing 12 or 24V DC systems. In a few cases, systems are configured at 48V DC. Methods of configuring systems are discussed in Chapter 9.

- Most small off-grid PV systems (especially solar home systems below 100Wp) use 12V DC as their system voltage. This means batteries are configured at

12V DC and the charge regulators and modules are rated at 12V DC. Lights are normally 12V DC in such small systems. If there is a need for AC power, an inverter is used to convert 12V DC electricity from the battery to the desired AC voltage.

- Sometimes 24 and 48V DC system voltage is used. In such cases, batteries and solar modules are wired in series or series-parallel so that they are 24 or 48V (see Figure 9.12, Chapter 9), and 24 or 48V charge regulators and inverters must be selected. Such systems have less voltage drop in wire runs, so they are often selected to save on cable costs (48V DC systems are common in off-grid telecom systems). However, note that 24 or 48V DC appliances are not readily available, so 12V DC system voltage is usually preferred.
- MPPT charge controllers accept electricity from the array at a range of voltages (i.e. from 15 to over 100V) and deliver it to the battery at 12 or 24V (see Chapter 5).
- Many charge regulators and inverters can operate at either 12 or 24V DC. They sense the system voltage and adjust to it automatically.
- Even though an inverter is used to convert power from DC to AC on the distribution side, the system voltage will still be between 12V and 48V.

Distribution Voltage(s)

This is the voltage of the circuits that carry power from the battery to the loads. At the planning stage, you must decide the distribution voltage of your loads:

- In small systems (e.g. SHS), most load circuits will be 12 or 24V DC.
- In systems with an inverter, some or all load circuits will be 110/230V AC.
- Some systems have circuits both in DC and AC (see examples in Chapter 11).

Box 8.2 Watt-hours and amp-hours

For planning purposes, energy consumption is indicated in watt-hours (Wh) or amp-hours (Ah). Watt-hours (or kilowatt-hours) are the most common measure of electric energy. However, because battery capacity and module output are measured in amp-hours, solar electric system planners often use amp-hours to indicate energy instead of watt-hours (strictly speaking, amp-hours are not a measure of energy, but a measure of charge). Make sure you know the difference between these two.

To calculate amp-hours, divide the energy in watt-hours by the system voltage:

Total charge (amp-hours) = energy (watt-hours)/system voltage (volts)

Example: A 12V system in a house with 4 lamps and an 80W television has an energy demand of 250 watt-hours per day. How many amp-hours does the system consume per day? Answer: Total charge in amp-hours = 250Wh/12V = 20.8Ah per day.

Worksheet 1: Total Daily System Energy Requirement and System Voltage

How much energy is needed to power a system each day? The steps below explain how to use Worksheet 1 to calculate the 'total daily system energy requirement'. This is the amount of energy the PV array must generate to meet the total daily load demand plus the extra energy required to cover system losses.

Before beginning, first make a list of all of the lights and appliances that the system will power. Shop around for low-voltage/low-energy appliances (especially lights) that you may want to use in the system.

Step 1: calculate the daily load energy demand in watt-hours

The 'daily load energy demand is the amount of energy required each day to power the load (i.e. lamps and appliances). This is measured in watt-hours and amp-hours. It is calculated by adding together the energy that all the individual appliances and lamps use on an average day. Separate the appliances into DC appliances (i.e. those that run at 12V DC or the system voltage) and AC appliances (those that will run through an inverter).

All loads are entered in the table and calculations of their energy use are made. Note that the top part of the table is for DC lights and appliances and the bottom part is for AC lights and appliances.

Column A: individual load description. List all the lamps and appliances to be powered by the system here. When listing appliances, you should consider all the appliances to be powered by the system, even those that will be purchased in the future.

Column B: individual lamp and appliance voltage. List the voltage of each of the appliances and lamps and whether they are AC or DC.

Column C: individual lamp and appliance power. List the power in watts of each appliance and lamp. Usually, the manufacturer indicates the power rating on the appliance itself. With better information about your appliance, you can accurately predict your demand. Try to get actual ratings of the power use of appliances from labels or manufacturers' data. Tables 6.1 and 6.3 (Chapter 6) list the power ratings of common lamps and appliances – if you do not have information about your appliances, use these when making calculations. Better yet, measure the actual DC current consumption of your lights and appliances using a multimeter or a clamp-on amp-meter. Do not guess! (The actual power consumption of AC appliances can be measured quite accurately using plug-in watt-meters, but it is best to use them on the mains – the wave form of an inverter may interfere with an accurate reading. Remember that in an off-grid PV system the appliance will consume more electricity because of inverter inefficiency.)

Column D: individual lamp and appliance use (hours per day). Estimate the number of hours per day that each lamp and appliance is used. If the appliance is only to be used a few times per week (e.g. a sewing machine might only be used on weekends), estimate the number of hours it is used per week, divide by 7 and write the number of hours per day in Column D.

Table 8.1 *Daily system energy requirement (from Worksheet 1)*

Column A **Lamp or appliance** List below	**Column B** **Voltage** Volts (V)	**Column C** **Power** Watts (W)	**Column D** **Daily use** Hours (h)	**Column E Daily energy use (DC)** Watt-hours (Wh)	**Column F Daily energy use (AC)** Watt-hours (Wh)
DC appliances					
Classroom 1 lighting (4 × 15W)	12	60	3	180	
Classroom 2 lighting (4 × 15W)	12	60	3	180	
Classroom 2 lighting (4 × 15W)	12	60	3	180	
Classroom 4 lighting (4 × 15W)	12	60	4	240	
Office Lights (2 × 15W)	12	30	1	30	
Staffroom lights (2 × 15W)	12	30	2	60	
Laptop computers (2 × 30W in lab)	12	60	3	180	
Admin laptop (1 × 30W)	12	30	4	120	
Security lights (2 × 15W)	12	30	6	180	
AC appliances					
Printer (1 × 20W)	240	20	0.5		10
TV/video (1 × 120W)	240	120	2		240
Cell phone recharging (4 × 5W)	240	20	6		120
BOX G: Total Daily DC Energy Demand				1350Wh	
BOX H: Total Daily AC Energy Demand					370Wh

2. Estimate system energy losses

Energy is always lost due to inefficiencies in cables, modules, batteries, charge controllers and inverters. The extra amount of energy lost must be estimated and added to the daily energy demand. For DC power multiply Box G by 0.20. For AC power multiply Box H by 0.35.

Box I: DC losses 270Wh

Box J: AC losses 130Wh

3. Add AC & DC Demand and AC & DC Losses

Box K: Total daily System Energy Requirement 2120Wh

4. Select System Voltage

Box L: System Voltage 12V

5. Calculate Daily System Charge Requirement

Divide Box K by the system voltage in Box L

Box M 177Ah

Column E: individual lamp and appliance energy use, DC (watt-hours per day). Multiply the power of each DC load (Column C) by the number of hours it is used per day (Column D) and write the figure in Column E. This is the energy use in watt-hours per day of each appliance.

Column F: individual lamp and appliance energy use, AC (watt-hours per day). Multiply the power of each AC load (Column C) by the number of hours it is used per day (Column D) and write the figure in Column F. This is the energy use in watt-hours per day of each AC appliance. Note that if you are not using AC appliances, leave this blank!

Box G: total daily DC load energy demand. Add the DC appliance totals in Column E, and write the total in Box G. This is the DC daily load energy demand in watt-hours.

Box H: total daily AC load energy demand. Add the AC appliance totals in Column F, and write the total in Box H. This is the AC daily load energy demand in watt-hours. Note that if you are not using an inverter, this should be zero.

Step 2: estimate system losses

Not all energy produced by the modules is available for use in the system, as some is lost in the cables, batteries, charge regulators and inverter. With the amp-hour method of planning systems (which uses the actual current outputs of modules in planning), there are concerns about two types of efficiency:

1. General System Efficiency: the efficiency of cables, the battery and the charge controller. This can be taken to be 80 per cent (meaning 20 per cent of the energy is lost).
2. Inverter Efficiency: a good inverter, one designed for use in off-grid PV systems for example, will be about 85 per cent efficient under average loads. Because of the extra losses in the inverter, AC energy in a system will suffer higher losses than DC energy, e.g. 80 per cent general efficiency multiplied by 85 per cent inverter efficiency equals about 65 per cent overall efficiency. This means that about 35 per cent of the electricity fed by the PV array into the battery that is converted into AC electricity by the inverter will be lost (the sum of battery and inverter efficiency).

Therefore, for DC energy losses (which don't pass through the inverter) multiply Box G by 20 per cent and write this number in Box I (DC losses). For AC energy losses, multiply Box J by 35 per cent and write this in Box J (AC losses). If you expect your inverter to be more (or less) efficient, you can change the value in Worksheet 1.

Step 3: calculate the total daily system energy requirement

Add the daily energy demand (Boxes G and H) to the estimated losses (Boxes I and J). This sum is the total daily system energy requirement in watt-hours. Enter this number in Box K of the worksheet.

Step 4: system voltage

Decide on the system voltage (see the System Voltage section, earlier in this chapter). Write the system voltage in Box L.

Step 5: daily system charge requirement

Divide the daily total system energy requirement (Box K) by the system voltage (Box L) to get the daily system charge requirement in amp-hours and write this in Box M. This is the charge in amp-hours that the module(s) will have to provide each day to meet the load requirements.

Box 8.3 Arusha farm system example

This example, taken from an actual case, is of a farmer/businessman from Arusha, Tanzania, who wishes to power the appliances in his off-grid residence: 15 lamps, a CD player, cell-phone rechargers, a laptop computer, a small-delivery water pump (not a borehole pump) and a colour TV (see Figure 8.2).

Column A **Lamp or Appliance** list below	**Column B** **Voltage** volts	**Column C** **Power** watts	**Column D** **Daily Use** hours	**Column E** **Daily Energy Use (DC)** watt hours	**Column F** **Daily Energy Use (AC)** watt hours
DC Appliances					
Lamps (6 x 3 W LED)	12	18	2	36	
Lamps (6 x 10W fluorescent)	12	60	2	120	
Sitting room lamps (2 x 15W fluor.)	12	30	3	90	
Security light (6W LED)	12	6	10	60	
Water pump	12	40	1.5	60	
				0	
AC Appliances		**154**			
Colour television	240	80	3		240
CD Player/music system	240	15	2		30
Laptop computer	240	25	3		75
Cell phone chargers (3 x 3W)	240	9	6		54
					0
		129			
BOX G: Total Daily DC Energy Demand				**366**	**Watt-hours**
BOX H: Total Daily AC Energy Demand					**399**

Figure 8.2 *Arusha total daily energy demand calculation (system and inverter losses not included)*

Step 1: Calculate the daily load energy demand for the appliances in the Arusha rural household: lamps, music system, TV, cell-phone chargers, a laptop and a water pump. These are listed in Column A with their voltage in Column B and their power demand in Column C. Note that the customer has selected efficient DC LED and fluorescent lamps, and that the pump is also DC (these are entered in the DC portion of the table). The laptop, TV, music system and phone chargers are 230V AC and written in the lower AC portion of the table. The daily use – in hours – of each appliance is estimated and listed in

Column D. For DC appliance daily energy use, Columns C and D are multiplied, written in Column E and totalled in Box G below (366Wh/day). Each AC appliance energy use is calculated in Column F and totalled in Box H (399 Wh). Altogether, the total daily energy use of all the appliances is expected to be about 765 Wh.

Step 2: Next, we need to estimate the system losses plus inverter losses. On the DC side of the system, losses are estimated as 20 per cent of 366Wh, or 73Wh (Box I). On the AC side of the system, losses are estimated as 35 per cent of 399Wh, or 140Wh (Box J). Total system losses will be about 213Wh per day!

Step 3: To get the total daily system energy requirement, add the load demand in Boxes G and H (366Wh plus 399Wh) to the losses in Boxes I and J (213Wh). This total, 978Wh, goes in Box K.

Step 4: As with most small PV units, system voltage in this case is 12 volts. All of the lights will run from a 12V circuit. The electrician will wire one 230V AC distribution circuit from the inverter to power sockets for the TV, CD player and laptop.

Step 5: Dividing the total energy requirement (978Wh, Box K) by the system voltage (12V, Box L) we get the daily system charge requirement of 81.5 amp-hours (written in Box M). This is the amount of charge that must be provided by the modules each day.

Worksheet 2: Sizing and Choosing the Module(s)

The size of the module or array is calculated using the daily total energy requirement and solar resource data for the site. To accurately calculate this, you will also need manufacturer's data (i.e. the I-V curve) of the solar modules being considered.

Step 1: calculate the solar insolation value for the site

Before selecting your module(s), you need to estimate the solar energy available at the site, i.e. mean daily insolation in peak sun hours per day or kilowatt-hours per square metre per day (kWh/m^2/day) in the design month. Records from a nearby meteorological station should be used if they are available (see Chapter 2). If there is no local meteorological station, estimate insolation using national records, which should be available for major cities (use the nearest city, see Chapter 12). Rough maps giving insolation data can be found in Chapter 2 and other sources of solar radiation data can be found on the internet (see Chapter 12).

First, from meteorological data, check the list of monthly mean daily insolation figures in peak sun hours or in kWh/m^2/day for a site near you. There should be 12 figures corresponding to the daily peak sun hours for each month; enter this monthly data in the table in Box 2A, Worksheet 2. Insolation ranges from 2 to 7 peak sun hours in various parts of the world.

Note that 'sunshine hours' are not the same as insolation or peak sun hours – they indicate how many hours the sun shines at the site, but they do not give the intensity of the sun! Do not use them. Neither should the average annual

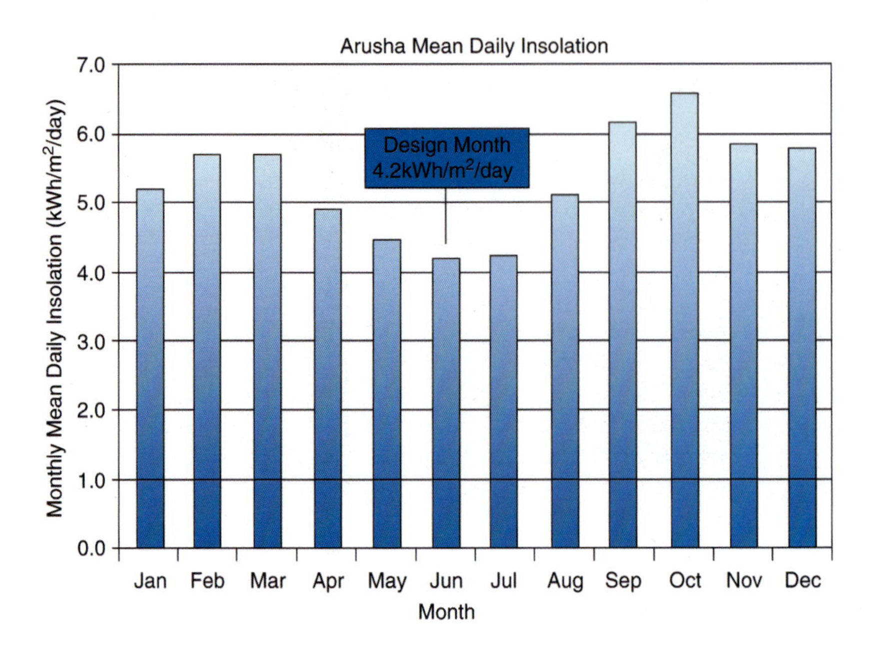

Figure 8.3 *Insolation data chart for solar resources*

mean daily be used. This is the average monthly daily insolation received over the year and will not take seasonal variations between summer and winter, or wet seasons and dry seasons, into consideration.

Next, check which month has the lowest mean daily insolation. This is called the 'design month' and it is used to size the array for stand-alone systems. Most off-grid PV systems are designed for the month of the year with the lowest insolation so that when the sun is least available, the system will work. However, there are exceptions – a pumping system may be designed to work best during a relatively sunny month and not during the least sunny month of the year when it is mostly cloudy but wet and so the requirement for pumped water is low.

Enter the design month insolation in Box 2B. This is the number of peak sun hours (or the total energy in kWh/m^2) expected per design day, from which the size of the array is calculated.

Finding insolation data

If no meteorological station is nearby, consult solar resource maps for your country to get a rough estimate of your site's annual mean daily insolation (see websites listed in Chapter 12), or ask a solar module salesperson what figure they use.

Compare your site's monthly weather variations with the monthly variations of sites nearest yours on the maps. Which month is the cloudiest, and how much more cloudy is it? In tropical and semi-tropical regions the design month usually has 15 to 25 per cent less insolation than the annual average. In extreme northern and southern latitudes, the design month insolation may be less than half the annual average.

Box 8.4 Design month and planning

A problem with using the 'design month' in planning is that, in some cases, the cost of the system will be greatly increased to cater for a low insolation design month. For example, a site with an overall annual daily insolation average of 5.5 peak sun hours might have a design month with 3.5PSH of insolation. As well as greatly increasing system costs, using the design month can result in the array producing too much power for much of the year.

If, for budgetary reasons, a smaller array is planned, then end-users must be aware of the lower electricity outputs and reduce energy use during low insolation months. Again, this sizing method recommends that planners always use the design month for system design. However, systems large enough to supply enough power all of the time are much more expensive than systems designed to supply electric energy requirements most of the time. (It is much less costly to have power needs met by solar 85 per cent of the time than 100 per cent of the time). If someone cannot afford the full size of the system initially needed, it is possible to start off with fewer modules and smaller loads.

Step 2: calculate the system design charging current

The module(s) in a system must be chosen so that their energy output matches the total daily system energy requirement. For your solar electric system to succeed over the long term, the average daily energy output of the modules must equal the average daily energy requirements.

In Worksheet 1, the system's daily total system charge requirement (in amp-hours) is calculated (Box M, Worksheet 1). Above, you determined the design solar insolation value at your site (Box 2B). Divide the charge requirement (Box 2C) by the design solar insolation value (Box 2D). Write the result in Box 2E. This is the 'system design charging current' – the charging current the array must produce to meet your energy requirements. You need enough modules to produce this current at your system voltage.

Step 3: determine the size of the module or number of modules required to produce the system design charging current

Now, select enough modules (i.e. a large enough array) to produce the system design charging current at the operating voltage and conditions at your site. To

Box 8.5 Daily system charge requirement

Energy Produced by Array per day (in amp-hours) = Daily System Charge Requirement (in amp-hours)

do this, divide the system design charging current by the normal operating current (see Box 8.6) output of the modules you have available.

Firstly, find out what modules are available and make sure you have the information sheets for the modules, i.e. manufacturers' tables or I-V curves. Use the current (in amps) that each module produces at the voltage of your system, and at the temperature and irradiance of your site. If you know the charging voltage and the operating temperature, you can use each module's I-V curve to calculate the operating current of the module. Remember, the higher the temperature, the lower the module output, and roof-mounted modules can be operating at 20°C (36°F) higher than the ambient temperature most of the time. Refer to Chapter 3 for more information about using I-V curves to determine module current.

As an example, the I-V curve in Figure 8.4 shows the output current of a module at a temperature of 47°C (117°F) charging a 12V battery (i.e. the ambient temperature is 27°C (81°F) and the modules are 47°C). Follow the dotted line up from 13 volts until it touches the 47°C curve, and then follow the dotted line across to the current reading (about 2.9 amps). Note that if the module temperature was 60°C (140°F), the current output would be significantly lower.

Table 8.2 shows the estimated current output and daily charge output of commonly available modules at three levels of insolation. (Note that the modules shown in this table are representative only; many other types are available.) As shown below, you can use the table to help choose your module(s). It shows module output at expected field temperatures and operating conditions – if module operating temperatures are lower than 50°C (122°F) at your site, your modules would have a higher output.

Example

Suppose the system design charging current required was 10.0 amps at 12 volts. None of the modules in Table 8.2 produce 10.0 amps. You would thus either need to find a larger module that produces 10.0 amps (i.e. a module of

Box 8.6 Normal module operating current

The 'normal module operating current' refers to the expected output current of a module at the operating temperature on a sunny day at the site where it is installed. Some module manufacturers provide data for expected module current output at normal operating cell temperature (NOCT), which is 47°C (117°F). This is a typical temperature for modules in hot locations, though expected temperatures may indeed be lower at your site. If your module manufacturer does not provide this data, use the I-V curves for the module at 40–50°C (104–122°F) to estimate its current output as explained below. Otherwise, consult the dealer or manufacturer for data. Table 8.2 provides estimated NOCT outputs of several modules.

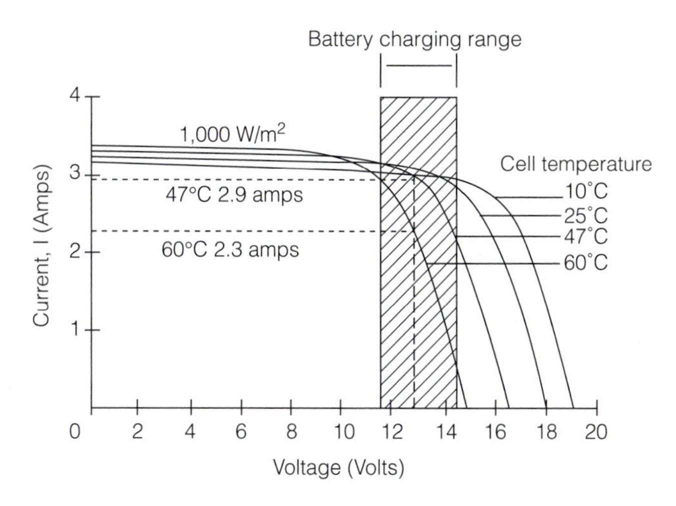

Figure 8.4 *Using I-V curves to calculate module current output*

180Wp or more), or buy several modules to meet this current demand. This would mean buying about four of the 40Wp amorphous modules, three of the 55Wp modules or two of the 80Wp modules.

Step 4: module voltage and configuration

Remember to select modules rated for your system voltage. Ask your supplier to help you ensure that you have the right voltage!

Table 8.2 *Approximate daily charge outputs of various 12V modules (47°C, 14V)*

Module Brand and Number	Type	Nominal Module Rating W	Module Operating Current Output (47°, 14V) Amps	Expected Daily Output @ 4psh Amp-hours	Expected Daily Output @ 5psh Amp-hours	Expected Daily Output @ 6psh Amp-hours
Kyocera KC40T	Polycrystalline	43	2.40	9.6	12	14.4
Kyocera KC50T	Polycrystalline	54	3.18	12.72	15.9	19.08
Kyocera KC85T	Polycrystalline	87	5.10	20.4	25.5	30.6
BPSolar BP 380	Polycrystalline	80	4.80	19.2	24	28.8
Sharp NE-80EJEA	Polycrystalline	80	5.00	20	25	30
Sharp ND-123UJF	Polycrystalline	123	7.80	31.2	39	46.8
Solengy SG Power 32-12	Amorphous	32	2.20	8.8	11	13.2
Solengy SG Power 10-12	Amorphous	10	0.75	3	3.75	4.5
Suntech STP020S-12/Cb	Polycrystalline	20	1.20	4.8	6	7.2
Suntech STP065-12/Sb	Polycrystalline	65	3.90	15.6	19.5	23.4
Kaneka P-LE055	Amorphous	55	3.60	14.4	18	21.6
Shurjo Energy SE25 H15	CIGS	25	1.80	7.2	9	10.8
Shurjo Energy SE37 H15	CIGS	37	2.50	10	12.5	15

Source: Manufacturers' data

- If your system voltage is 12V DC, select 12V-rated modules (with 36-cells).
- If your system voltage is 24V DC, select two 12V-rated modules and arrange them in series (or use 72-cell modules rated for 24V). You will need to select double the number of modules required to reach the charging current.
- If using a MPPT charge controller follow the instructions in the charge controller manual. MPPTs can accept a large range of voltages, as well as increasing module output. The design method being used here assumes that there is no MPPT in the system (with an MPPT charger, expected module output would be considerably higher). See Chapter 11 for more details.

Box 8.7 Arusha farm system example: sizing and selecting modules

In this example, we use meteorological data to calculate how many modules will be required to power the load (already calculated) for the Arusha system.

Step 1: first find insolation values for the site near Arusha (this is available from Tanzanian meteorological offices). Values for Arusha are entered in Box 2A as shown in Figure 8.5.

Box 2A Arusha Solar Records

Mean insolation data should be filled below.

Month	Peak Sun Hours
Jan	5.2
Feb	5.7
Mar	5.7
Apr	4.9
May	4.5
Jun	4.2
Jul	4.3
Aug	5.1
Sep	6.1
Oct	6.6
Nov	5.9
Dec	5.8

Figure 8.5 *Arusha insolation data entered in Worksheet 2*

Step 2: as shown, the lowest monthly value is 4.2kWh/m^2 in June (the annual daily average is about 5.3kWh/m^2). This is the design solar insolation value and is entered in Box 2B.

Step 3: the system design charging current is calculated by dividing the daily system charge requirement (81.5 amp-hours, from Worksheet 1, Box M) in Box 2C by the design solar insolation value (4.2 kWh/m^2) in Box 2D. The result, 19.4 amps, is the charge the array must produce under normal operating conditions.

Step 4: to find the most appropriate modules, shop around, consult solar dealers and get their datasheets. From the modules listed on Table 8.2, the following module configurations are possible for the Arusha system:

- Three of the Sharp 123Wp modules. (This system would be oversized, as it would produce 23.4A!)
- Four of the Kyocera 85Wp, BP Solar 80Wp or Sharp 80Wp modules.
- Five of the Suntech 65Wp or Kaneka 55Wp modules. (The Kaneka system would be slightly small at 18A.)

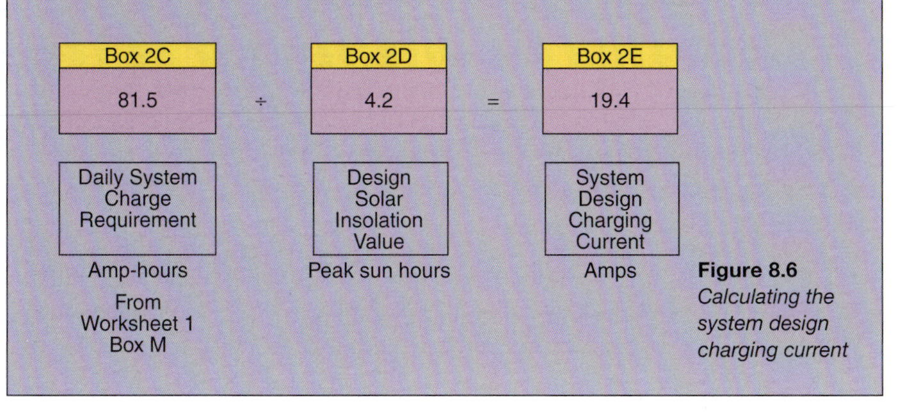

Figure 8.6 *Calculating the system design charging current*

Worksheet 3: Battery Sizing and Selection

Choosing the battery

As the part of the system most likely to experience problems, the battery must be selected and sized carefully. Worksheet 3 guides you through battery sizing calculations.

Choice of battery will be limited by what is on the market and how much you have to spend. Consult Chapter 4 for more information on battery selection. Remember, buy a good battery if you can afford it – it will pay off over the years. Do not choose automotive or second-hand batteries (choose modified SLI or truck batteries instead!).

The following steps take you through the process of choosing and sizing your batteries:

Step 1: identify the available batteries in your region that fit within your budget

Make a list of these batteries and determine which type (see also Chapter 4 for more information about batteries). Do a table similar to Table 8.3 below and make a comprehensive list of available batteries, listing all their important characteristics. Be careful if your supplier cannot provide this basic information about their batteries! If in doubt, check on the internet or elsewhere (see Chapter 12 for more information).

Once you have a table with information about the types of batteries available, consider which type you think is best. Cost per cycle is a great way to compare (divide the price by cycle life to get this). Select the best two or three options and use each in the following calculations before making the final choice.

Step 2: calculate the required capacity of your battery (in amp-hours)

The capacity of the battery required depends on three primary factors:

- The daily system charge requirement in amp-hours that must be supplied to the loads each day. This has already been calculated in Worksheet 1 (Box M).
- The maximum allowable depth of discharge factor. This is the deepest depth of discharge that is ordinarily allowed with the battery. Shallow cycle batteries, for example, should not be cycled below 20 per cent depth of discharge, while deep discharge batteries can regularly handle 50 per cent discharges (see Chapter 4).
- The reserve storage factor, or the number of days of storage needed. This varies with site and is higher for sites with cloudy weather. In sunny areas, depending on the application, this number may be as low as one to two

Table 8.3 *Know the features of available batteries (fill in the table with information from suppliers for batteries you find)*

Type of Battery	Voltage	Capacity (Ah)	Recommended Daily DoD (%)	Recommended Max DoD (%)	Price	Cycle Life @ 25% DoD
Modified SLI (Portable) Make:						
Traction Battery Make:						
Captive Electrolyte Gel Make:						
Absorbed Glass Mat Make:						
Tubular Plate Batteries Make:						
Other						

days only. Do not use a reserve of greater than four days because it will greatly increase the cost of the battery bank and it will increase the risk that, during cloudy weather or winter periods, batteries will not be fully charged and will be damaged by cycling in a low state of charge.

To complete the calculations, enter the daily system charge requirement in Box 3A of Worksheet 3. Enter the reserve days (a number between 1 and 4) in Box 3B. Enter the maximum daily depth of discharge from the battery manufacturer in Box 3C as a decimal. (Note not to use a number over 0.8 as deep discharges over 80 per cent will damage even a good battery). The required system battery capacity can now be calculated by completing the equation and writing the answer in Box 3D.

Remember, over-sizing the battery is not recommended, especially where there are long cloudy periods and where load management is not strict. Most system designers compromise between cost and reliability by selecting a battery with a capacity of between two (in very sunny locations) and five (locations with cloudy periods) times the daily system charge requirement. If appliances are mostly used during the day (when the sun is shining) then the need for battery capacity is reduced (some PV simulation software can model battery size using this parameter, see Chapter 12).

Step 3: determine the configuration of your battery set

Batteries should be configured at the system voltage. Make a drawing of your battery configuration. Get advice if you are unsure how to do this.

- Small off-grid PV systems usually use 12V DC as their system voltage and 12V batteries. If more than one 12V battery is used, they are configured in parallel.
- If 2V DC or 6V DC batteries are used, they are configured in series so that their total equals the system voltage (i.e. at a system voltage of 12V DC, two 6V batteries would be configured in parallel).

Step 4: determine if your battery needs a special charge controller setting

Sealed AGM and gel cell batteries require a special setting on the charge controller. Make sure you know the charge regulation requirements of your battery.

Box 8.8 Calculating capacity of required system battery

Daily System Charge Requirement ×	Reserve Days ÷	Maximum DoD =	Capacity of Required System Battery
Ah	Days	%	Ah

Step 5: know your battery's maintenance requirements and cycle life

No matter what type of battery you choose, it is not something that you should install and forget! Virtually all wet cell batteries need to be cleaned and topped up with distilled water regularly. Make sure that system owners know how to do this (see Chapter 10). Also, make sure you know the how long the battery is likely to last (remember, the cycle life is about equal to the life in days) so that the system owner knows approximately when it will need to be replaced.

Box 8.9 Arusha battery sizing example

Using Worksheet 3, we can decide on the battery type, capacity and configuration for the Arusha system.

Step 1: a few scouting visits to the few battery suppliers in Arusha town reveals that (aside from automotive batteries which will not be used) the battery types available are:

- 100Ah 12V modified SLI batteries;
- 350Ah 6V traction batteries;
- 200Ah 12V AGM batteries (see Table 8.4 for details.)

Table 8.4 *Batteries available in Arusha*

Type of battery	Voltage	Capacity	Recommended daily DoD (%)	Recommended Max DoD (%)	Price (USD)	Cycle life @ 25% DoD
Modified SLI (Portable)	12	100Ah	20%	30%	$120	400
Traction Battery	6	350Ah	30%	50%	$400	700
Absorbed Glass Mat	12	200Ah	30%	50%	$450	1200

Step 2: using the worksheet, the required battery capacity is estimated by multiplying the daily system charge requirement by the reserve days (for Arusha it is decided to use 2 days) and dividing this by the Maximum DoD of the battery type selected.

- If the 100Ah modified SLI is selected, 543Ah capacity would be needed. This would mean purchasing five or six batteries and connecting them in parallel.
- If the 6V 350Ah traction batteries are used, a battery capacity of 326Ah would be required. This would mean that two batteries would have to be utilized and connected in series to reach 12V.
- If the 200Ah AGM batteries are selected, a capacity of 326Ah would be required. This would mean purchasing two of the AGM batteries and connecting them in parallel.

From the above, the traction or AGM battery appears to be the best value (five of the modified SLI batteries would have a shorter life)!

Step 3: as mentioned above, two 350Ah 6V traction batteries would be connected in series to reach 350Ah at 12V.

Step 4: the traction batteries, which are wet cells, use normal charge control settings. They will require occasional equalization charges.

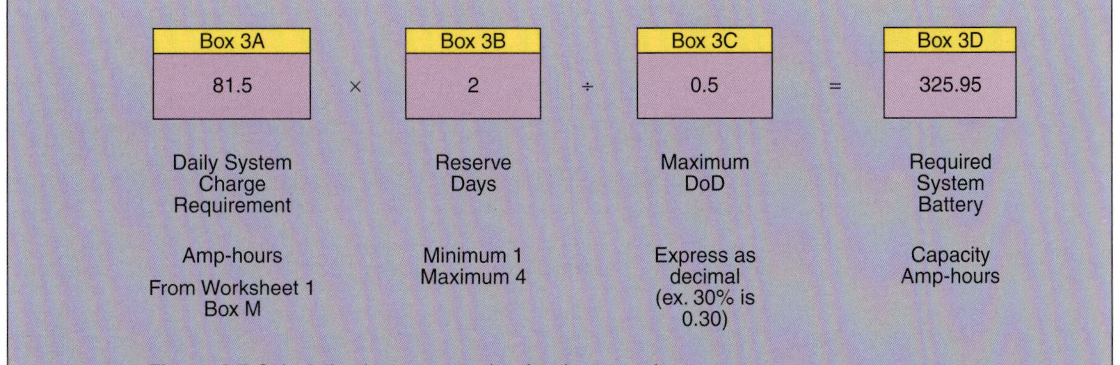

Figure 8.7 *Calculating battery capacity: Arusha example*

Step 5: based on manufacturer's specifications, the batteries should have a life of 700–800 cycles, or about 2–3 years. With better treatment, they can last longer. The batteries will need to be topped up on a monthly basis and put though an equalization charge quarterly.

Step 6: in this system the expected daily energy use is about 765Wh: about 64Ah at 12V. Over a 24-hour period this converts into a constant discharge current of about 5 amps but the system is unlikely to be powering appliances for 24 hours – an 8-hour 'on' period is probably more realistic. That would give a discharge current of about 15 amps. So, if the battery 350Ah capacity is for C100, the actual capacity of the battery would be less; but if the 350Ah capacity is for C20, that would be more accurate – taking 15 amps from the battery would empty it in about 23 hours. This will affect battery cycle life. See Chapter 4 for a discussion of battery capacities at different rates of discharge/C-rates.

Step 7: the maximum discharge current to which the battery will be subjected (when all DC are on and the inverter is working at full load, taking into account inverter inefficiency) needs to be calculated. You need to check the battery specifications or with supplier whether this is acceptable. It usually is, but this can be an issue in systems with large inverters.

Worksheet 4: Choosing the Charge Controller/Inverter Sizing and Selection

Charge Controllers

All off-grid solar PV systems need some type of charge controller and there are various types that offer a range of functions. Read Chapter 5 carefully to learn about charge controllers.

Before choosing a charge controller, the planner must:

- decide what controller size (rating in amps) is needed;
- decide what controller features are required.

Controllers need to be sized to the system voltage – if the system is 12V you need a 12V controller and if the system is 24V, you need a 24V controller. Worksheet 4 provides space to answer these questions.

Step 1: select the charge controller size

The controller must be sized to handle both the maximum short-circuit current from the array and the maximum demand of the load. Charge controller sizes range from 5 to 50 amps or larger. Note that many charge controllers have the same 'charge' and 'load output' rating (i.e. a 10A controller is sized to accept 10A of charge from the array and to provide a maximum of 10A load output).

- To determine the array current size, calculate the maximum short-circuit current of the array (Box 4A) and multiply it by 1.25.
- To determine the maximum DC load output current, calculate the maximum power consumption of all DC (not AC) appliances (see Worksheet 1, Column C). Divide this by the system voltage to get the maximum DC load. Write this number in Box 4C and multiply by 1.25. Write the result in Box 4D. This is the load output current rating in amps.

Step 2: select desired charge controller features

Depending on the size of the system, the appliances used and the solar insolation at the site, a controller will be required to perform different functions. Chapter 5 discusses the features of charge controllers, and Table 5.1 shows the features of common controllers. Important features to consider include: overcharge protection, low-voltage disconnect, solar charge and low-voltage warning lamps, voltage and current meters, and load timers. Fill out the 'Desired Features' tables in Worksheet 4 to get an idea of the charge controller you need.

- Features: low-voltage disconnect (LVD) and high-voltage disconnect (HVD) are important features of charge controllers. Make sure you know the rated values of the unit you want to purchase.
- Displays: simple charge controllers use one or more diodes to inform users about solar charge and state of charge. More elaborate charge controllers have liquid crystal displays of the amp-hour meter. Select the display appropriate to customer needs.
- Protective features: the better the controller, the more protective features it has. Check which type of short-circuit and reverse polarity protection it has.
- Charging type: be sure that the charging method your controller uses matches the needs of your modules and batteries.

Inverters

Some off-grid solar PV systems require inverters. There are various types that offer a range of functions. Read Chapter 5 carefully to learn about inverters.

Because of the wide ranges of inverter types and prices, it is a good idea to carefully consider what you need and what you can afford. Remember, most inverters are not meant for solar PV systems and can end up causing problems because of poor efficiency and wave-shapes! See Chapter 5 (and the resource section on inverters in Chapter 12) for more information.

Step 1: is an inverter needed?

Many small off-grid PV systems operate 100 per cent on 12V DC electricity and do not require inverters. If the system is below 100Wp and you can procure all of your appliances in 12V, a 12V DC system without an inverter may be the best choice.

Step 2: do you need an inverter-charger?

If an inverter is needed, decide whether it will function only as an inverter (i.e. DC to AC only) or if it also needs an integrated battery-charger. Inverter-chargers (usually for systems above 500Wp only) accept power from generators (or the mains grid in the case of back-up systems) and can charge the batteries when the solar resource is low. But remember, these sizing worksheets are not designed for large hybrid systems.

Step 3: inverter rating

The inverter must deliver the maximum AC load that you expect from AC devices. Total up the power of the AC devices in Column C, Worksheet 1, and write this total in Box 4E. Multiply this by 1.25. Write the answer in Box 4F: this is the recommended inverter rating.

Step 4: select desired inverter features

Review Chapter 5 and fill in the table with the desired features of the inverter you need. Make sure the wave-shape of the inverter matches the needs of your appliances.

Box 8.10 Arusha system example: charge controller and inverter selection

The Charge Controller

Step 1: to size the charge controller, the short-circuit current (Isc) of the array is calculated and multiplied by 1.25. In this example, if four Sharp modules were selected, then the Isc of one 80Wp module (5.15A) would be multiplied by four to get the array Isc (20.6A) and this would be multiplied by 1.25. The minimum size for the regulator would thus be 25.8A. Always read charge-controller technical information and instructions, as they may have more information about sizing.

Add up the power ratings of all the DC loads (Worksheet 1, DC section of Column C) and divide it by the system voltage to calculate the Maximum DC load. In this case, divide 154W by 12V and multiply the product by 1.25 to get 16A.

The charge regulator will normally be based on the larger of the two above.

Step 2: in selecting the charge controller features for an elaborate system like this, it is necessary to make sure that the controller does enough to prevent faults (e.g. short circuits) and also to let the consumer know that the system components are working. See Figure 8.8 for the desired features of this system.

Controller Specification		Rating
Rated Voltage	V	12
Maximum Array ISC Input	A	30
Maximum Load Output	A	30
Self Consumption	mA	

Feature	YES/NO	
High voltage disconnect	YES	14.5V
Low voltage disconnect	YES	11.5V
Temperature compensation	NO	
Load timer	NO	

Protection	YES/NO
Short circuit protection (array)	YES
Short circuit protection (load)	YES
Reverse polarity (array)	YES
Reverse polarity (load & battery)	YES
Lightning protection	YES
Open circuit battery	YES
Sealed battery charge settings	NO

Displays		YES/NO
Solar charge indicator		YES
SoC indicator (LED)		NO
LCD display		YES
Amp hour meter		YES

Desired Charging Type	YES/NO
Pulse Width Modulation	YES
Equalization	YES
MPPT	NO
Other	

Figure 8.8 *Choosing a charge controller*

Note that:

- High- and low-voltage disconnects are important in large systems.
- In an area with a long cloudy season, like Arusha, it is important to have a good LED display on the charge controller. Many 30A units also incorporate amp-hour meters into their circuitry.

The Inverter

Step 1 and 2: since there are a number of AC appliances in the Arusha system (TV, laptop, CD player and cell-phone chargers), an inverter is required for this system. However, since PV will be the only charging source, an inverter-charger is not required.

Step 3: to calculate the size, tally all of the AC loads from Worksheet 1, Column C (129W), and multiply by 1.25 to provide a margin. From this calculation, a 161W inverter is required. Small inverters come in a variety of sizes – it would be possible to find a 200W or 250W unit.

Step 4: when choosing the features of the inverter, consider the types of appliances that will be used. The peak surge power is not important in this case as the TV, laptop and music system appliances do not cause a demand surge. However, because the sound equipment used has sensitive motors and does not tolerate electrical interference, it would be wise to select a sine-wave inverter in this case.

Worksheet 5: Choosing Cables and Fixtures

Use Worksheet 5 (from the downloadable spreadsheet) to determine the size of the cables and fuses required in your system. The process of selecting properly sized cables is carefully explained in Chapter 9 and needs to be referred to. The steps below provide an order for you to follow as you complete the design process. Once you have completed this section, you should be able to draw up a list of all the materials you need to complete the installation. For a discussion of earthing and lightning protection requirements see Chapter 7.

Step 1: make floorplans or drawings of the site

Draw a floorplan of the house or institution in the space provided (or on a larger piece of paper if the space is not enough). This drawing should be to scale if possible, so that you can estimate cable lengths. Indicate clearly where each lamp, socket and switch will be, and the position of the battery, control and module (refer to and read the appropriate chapters to decide on the placement of these items).

Step 2: determine the correct size of cables for each run to avoid voltage drop

See Chapter 7 for the cable sizing process. Use the cable runs and estimated loads to determine the necessary cable sizes. Do this using Worksheet 5 (see also Table 7.4).

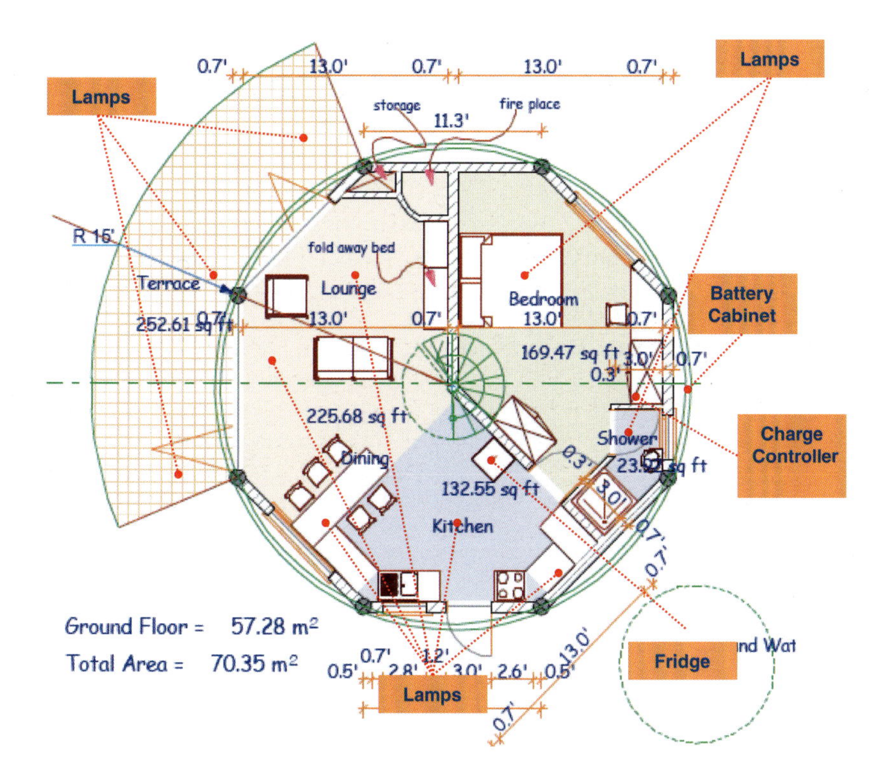

Figure 8.9 *Using floorplans*

Usually, the run from the module(s) to the control, and from the control to the battery, requires a larger wire size. Your system will get the most energy if you size battery and module cables for a voltage drop of less than 2 per cent.

Determine the total length of each type of cable required. Measure the distance of each run between the loads and battery, and between the battery and the module, remembering that cables go up and down, as well as across. For remote installations, it is always good to overestimate the amount of cable required as trips to get more cable are expensive.

Step 3: size fuses for each major circuit

Follow the instructions in Worksheet 5, and refer to Chapter 7 as necessary. Remember, it is crucial for the system's safety that you put a fuse between the battery and the controller, and between each major load and the controller.

Step 4: determine the type and quantity of wiring accessories

These accessories include sockets, fuses, switches, connector strips and clips, tacks, etc. Make a list of the materials required before going to purchase them.

You have now finished planning for the wiring of the system. Before purchasing anything, you should make a list of all of the materials you need, down to the last tack, so that you can tick it off when you are making your purchases (see Table 8.5).

Table 8.5 *Arusha planning example*

Component	Size	Description	Numbers/Amount
Solar module	80Wp	Monocrystalline	4
Battery	350Ah	6V Traction	2
Charge controller	30A	PWM with LVD	1
Inverter	250W	Sine wave	1
AC circuit cables	2.5mm^2	Twin flat	30m
DC circuit cables (all sizes and types)	• 2.5mm^2 • 4.0mm^2 • 6.0mm^2	Twin flat, multi-strand cable	• 80m • 30m • 20m
Conduit	standard	For exposed cables	30m
Switches	5A	DC rated switches	20
Sockets	240V AC, 5A	Switched	4
Fuses	50A	Main battery fuse DC rated	1
Junction boxes	Standard DC		40
Connector strips	Standard DC	Standard	4 boxes
Earthing		Earth rod	1
Bolts, screws, nuts, etc.		Various boxes	

9

Installing Solar Electric Systems

This chapter describes the process of installing solar electric systems. Within the scope of this book, it is impossible to cover all aspects of electrical installation. This chapter has therefore been written mainly to guide readers through the installation of small solar PV systems with 12 or 24V DC circuits. Nevertheless, much of what is covered herein is also applicable to other types of PV systems. Readers wishing to design and install larger and more complex systems should undertake general electrical training as well as training in off-grid PV system design and installation (see Chapter 12).

The chapter guides the reader through the following tasks:

- Preparation of tools and materials necessary to complete an installation.
- Positioning, mounting and wiring solar cell modules.
- Wiring the control and batteries.
- Laying the cables.
- Making and inspecting the final connections.
- Commissioning the system and training the system managers.

Foreword to Installation

Once a system has been planned and the necessary equipment obtained, installation can begin. In all cases, a competent person should supervise work to ensure that the system is installed according to the local electric safety codes and so that no person is injured or equipment damaged during installation.

The installation methods presented here are based on techniques that have been field-tested. Hundreds of thousands of small systems have been installed worldwide using these techniques since the mid-1980s. In general, the methods are similar to wiring practice already familiar to electricians, although low-voltage DC wiring has some important differences from AC wiring.

Note that in many countries standards and codes of practices exist for solar electric installations. They provide guidelines for installing systems that should be followed. In some countries, solar electric system owners and installers may be legally responsible for upholding codes of practice. Check with solar companies in your country to find out if standards or guidelines are in force. See also Chapter 12 for lists of downloadable international standards.

Box 9.1 Suggested installation procedure

The sequence will differ from site to site and from installer to installer. It is important to plan the job, and follow a logical plan:

1. Check to make sure that all equipment is on-site.
2. Finalize the choice of locations for the solar modules, controller, inverters and batteries.
3. Lay cables to and install PV modules.
4. Lay cables for and install batteries.
5. Lay cables for and install charge controller.
6. Lay cables to loads. Attach lamps and sockets.
7. Complete the final connection sequence and commission system.

Tasks need not necessarily be carried out in this order – it depends on the installation. For example, you may want to install the PV modules first and start charging the battery while the distribution circuits are being installed.

Installation Tips

- Before beginning an installation, make sure that all equipment is at hand. This includes tools, materials, necessary spares and information resources. Many solar electric installations are conducted in remote areas where equipment and spares are not available. Delays are expensive and they often occur because basic spares like cables, connector strips or special bolts went missing!
- Read the manuals for all components. Follow their directions carefully.
- Follow the recommended sequence of installation. Do not connect appliances, lamps, batteries or solar cell modules to the controller, until the last step. Follow the final connection sequence carefully (see below).
- If you are not experienced with electrical installations, complete the installation with help from a competent expert.
- Always use the proper tools for each task (see next section).
- Maintain high work standards. Work standards refer to the way the wires are laid, the consistency of switch placement, the method with which fixtures are attached to walls and the general neatness of the work. High standards will make the system look more attractive and last longer, and will add to the system safety.

Tools, Instruments and Materials

It may be difficult to obtain tools, extra parts and equipment while on-site. For this reason – before departing for the installation site – make check lists of all the materials and tools needed. This list should be carefully cross-checked during trip preparations. Use the information gathered during the planning stages (i.e. the map of the site, the worksheets and the circuit diagram, see Chapter 8), together with Tables 9.1 and 10.2, to make the checklists.

Unless you bring a portable generator, there will not be 230V power for tools. You will require power tools rated at 12V DC (or hand tools). If using 12V DC tools, fill and charge a battery upon arrival at the site, so that there is power for tools and lights during the installation.

The installation, maintenance and troubleshooting work described in this book require that installers use a digital multimeter. Multimeters are essential when checking continuity (broken wires, bad connections), insulation, resistance and polarity, and when measuring the voltage of modules and batteries. Digital multimeters also allow measurement of current, which is useful when installing and testing modules. Auto-ranging multimeters are more expensive but more convenient to use. However, it is not the only instrument used in off-grid systems – insulation testers and other instruments are also used (see below).

Installers need to be fully familiar with the use of a digital multimeter, as a mistake may damage the meter or cause injury (consult the manual). Note that most multimeters measure DC current using separate lead outlets for current and volts: make sure you always change the lead! Also, current measurement

Table 9.1 *Basic tools for a solar electric installation*

Tool	Purpose of tool
Crimping tool (see page 149)	Attaching bootlace, ring and spade terminals to wires
12V DC Soldering iron	Connecting wires to terminals, fixing electrical parts
Digital multimeter	Testing connections, measuring voltage, needs to be able to measure DC current up to 10A at least, and be properly fused
Screw drivers (star- and flat-bladed, insulated)	Tightening screws and terminals
Hydrometer	Measuring battery state of charge
12V drill and drill bits (if not available, use a hand drill)	Drilling holes for various purposes
Tape measure	Measuring distances and marking wire clip placement
Pencil and paper	Taking notes on measurements
Hack saw	Cutting metal frames
Utility knife	Various cutting jobs
Wire cutter and stripper	Preparing cables
Torch	Laying wires in dark places (ceiling), working after dark
Pliers	Holding bolts and nuts during tightening
Adjustable spanner	Tightening battery terminals
Hammer	Various construction tasks
Shovel	Digging trenches, foundations
Level	Checking grade of mount, laid wire and foundations
File	Smoothing rough surfaces after cutting
Extension cord	Running power from inverters to tools
Inclinometer and compass	Fixing the angle of solar modules
Product literature for system components	Source of reference information

Figure 9.1 *Digital multimeters*

usually is limited to a maximum of either 2 or 10 amps. Because some measurements can exceed 10 amps (e.g. battery current, large array current) make sure that the right measurement settings are always selected and that the meter is fused (for current measurements). For those doing many large installations, clamp-on amp-meters are a better choice, as they can measure current without requiring disconnection of the cable (see Chapter 11 for information about tools for large systems).

Safety

Installing small solar PV systems is not considered particularly hazardous, but all electrical and building work can be dangerous if safe working practice is not observed. Solar equipment and batteries have the potential to cause serious injury. In the remote areas where so many systems are installed, it is especially important to use care and avoid accidents. Remember, PV installations have two power sources – the battery and the modules – and these are both potentially dangerous if you do not take precautions.

In general, to avoid shock risks when working with electricity:

- always remove rings and jewellery;
- use insulated tools;
- keep loose cables and metal tools away from the controller, batteries and arrays so that they do not accidentally come into contact with live terminals or leads;
- always be aware of possible shocks from modules and batteries, and take steps to avoid them;
- tape up all cable ends (separating all conductors from each other in the case of two or three-core cables) with insulation tape when cables are hanging loose (not fixed into terminals); the ends of all live cables (cables coming from batteries and array) should be taped as a matter of course but also non-live cables (they can become energized by accident);

Box 9.2 Battery safety

When working on batteries:

- remove tool belts;
- remove jewellery;
- use insulated tools;
- use the correct tool for the job;
- handle tools carefully;
- put tools on the floor when not using them;
- wear goggles when topping batteries with acid; and
- always have plenty of water on hand.

- remove main fuses (especially the battery fuse) when working on the system and put them in your pocket so that nobody else can replace them while work is being done on circuits;
- do not work on live extra-low 12/24V DC voltage circuits unless absolutely necessary and you know exactly what you are doing – and never work on live 110/230V AC voltage circuits under any circumstances;
- when wiring modules beware that they are a source of voltage – and if connected incorrectly can produce high voltages – so special care needs to be taken; one safety measure is to cover the array with a blanket until the time comes to connect the PV module(s) to the charge controller.

Batteries need to be handled with special care (see also Chapter 4):

- Battery acid is extremely corrosive. It can destroy clothes, burn skin or cause blindness if it comes in contact with the eyes. Wear protective clothing and glasses, and use a funnel to avoid splashing when filling cells. Always keep fresh water available to rinse spilled acid from clothes, hands and eyes. Baking powder neutralizes acid spilled on clothes and on the floor.
- Batteries are heavy. Carry them upright, from the bottom, or by the handles provided. Never lift batteries by the terminals.
- Make sure that batteries are located in a ventilated space. Do not smoke near batteries.
- Beware of the electrical charge in a battery. If the terminals of the battery are accidentally shorted, there is a possibility of explosion or electrical shock.
- Be extra careful when transporting charged batteries. Cover the terminals and make sure nothing is placed on top of them.
- Work on large battery banks should only be done by persons who have received appropriate training.

Modules are expensive and generally robust but can be damaged, so:

- Transport with care. Be careful of the back of the module which is especially fragile.

- Beware of shocks when wiring and installing modules. Although a 50Wp module cannot easily cause a lethal shock, several modules in series or parallel are more dangerous. If someone receives a shock from a module while working on a roof, the shock itself may not be very serious but it could cause a fall and serious injury or death.
- Disconnect or cover the array with a blanket when wiring to avoid current flow and electric shocks.

Ladders/roof mounting: a large percentage of injuries incurred when installing solar electric installations are caused by falls from the roof. When on the roof:

- Use stable ladders, and position them correctly. Have somebody hold the bottom part of the ladder.
- Use ladders or planks to move about on the roof, and scaffolding as necessary. Safety harnesses are also available.
- Wear protective headgear when handing large modules and mounting structures.

First aid kits and plenty of water should be on the site during any installation at all times:

- Keep the first aid kit well-stocked and ready.
- Keep water in a place where everyone can quickly reach it.
- Make sure someone knows how to use the first aid kit.

Laying Cables

Laying cables is usually the first task during a solar electric installation. If the job will take several days, lamps and tools can be temporarily connected so that technicians have a well-lit, convenient place to organize and conduct assembly work even after sunset. If the modules are to be connected at a later stage, start charging batteries immediately to ensure there is power on-site, and to ensure that the batteries will have a full charge when the installation is complete.

Wiring Guidelines for Small Off-Grid PV Systems

Wiring should always be carried out to the standards in national electrical codes and regulations. Since most national codes are primarily written for grid electricity, PV may not be covered. If your country has national codes for off-grid PV or solar home systems, follow them.

The following guidelines are provided for those installing small solar home systems up to about 500Wp with 12V or 24V DC distribution circuits for lights and appliances:

Leave a circuit diagram

Put a copy of the diagram near the control box. The more complex a system is, the more vital a circuit diagram is as a reference for electricians repairing the

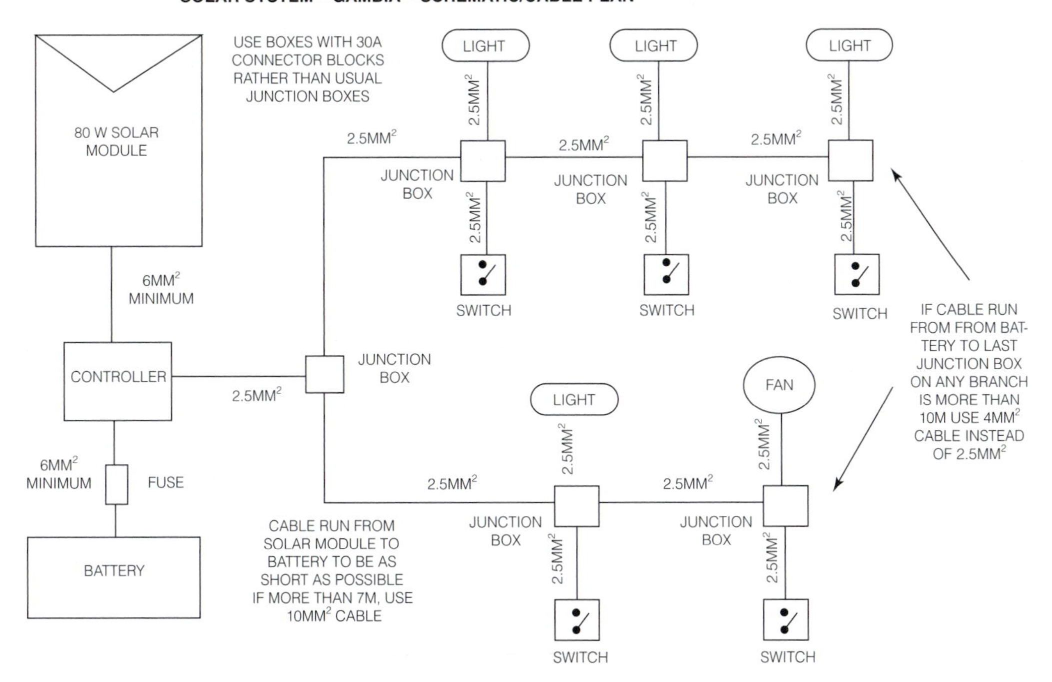

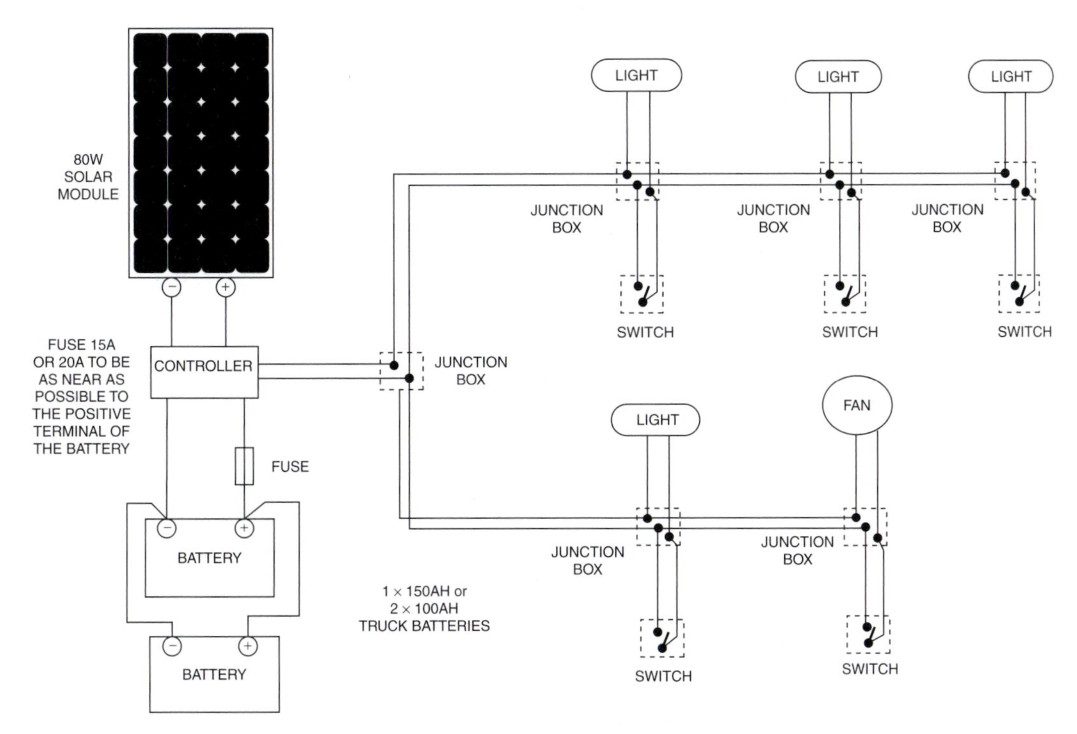

Figure 9.2 *Cable plans and wiring diagrams are used by electricians to plan wire layout*

Source: Green Dragon Energy, www.greendragonenergy.co.uk

system in the future. Circuit diagrams help enable electricians to quickly understand how a system is wired when they do expansions or repairs.

Follow established cable colour codes

Electrical cable is insulated with colour-coded covering which specifies its polarity. Systems that do not use proper wire-colouring are hazardous and can be illegal. Some established DC wire colouring codes are given in Table 7.2. In systems where existing wiring is being upgraded or replaced, be aware that different colour codes may have been used – cable colour codes vary in different countries and some may have recently changed!

Label cables during work

When running cables, label them on both ends with electrical tape for quick polarity identification. When working with a cluster of cables passing through a wall or conduit, labelling will avoid wasting time checking which cable connects to which source. For example, a positive cable from the charge controller to the array should be labelled 'PV ARRAY +' so that it can be identified even if it accidentally gets disconnected.

Always double check polarity when wiring

Accidentally fixing the wrong cable to a terminal can be disastrous. Some equipment can be damaged by reversed polarity. Do not connect lamp fittings and other items of equipment until DC polarity has been checked.

Earth module frames and loads

If required (or necessary), earth systems for safety and for protection from lightning (follow guidelines in Chapter 7).

12/24V DC Wiring Practice

Voltage drop will occur wherever there are poor connections, so make them as tight and secure as possible (see Chapter 7 for a discussion of cable types).

Preparing cables

- Cut the exact amount of insulation required from the end of the wire.
- Avoid cutting or breaking wire strands.
- Use crimp-connectors if at all possible (otherwise, twist the wire strands before inserting them into connector strips).

Laying cables

- Use correctly sized connector strips for joining cables.
- Never make 'twist' connections between cables as they are likely to come apart or cause a voltage drop.

- When tightening screws in connector strips and terminals, turn the screw until it is tight enough for a secure connection, but not so tight that it cuts the wire.
- All connections should be enclosed in the junction/connector box and be in reasonably accessible locations so that they can be accessed if there is a problem.
- Clip cables neatly to the wall, spacing the clips at regular intervals, or run cables in a conduit that is properly fixed to the wall.
- Make sure all wire runs are horizontal or vertical (use a levelling tool).
- Use a crimping tool if one is available. Crimping tools and connectors make secure electrical contacts and enable work to be done quickly (see Box 9.3).
- When laying a conduit outdoors, make sure it is supported every 3m (10 feet) and within 6cm (2.5 inches) of the electrical box.
- When running a wire from one building to another it should be supported by a catenary wire.

Situating switches, sockets and junction boxes

- Light switches and power sockets should be in wall-mounted pattress boxes (i.e. back boxes).
- Consider the needs of users and local codes.
- Use standard 'off' and 'on' switch positions throughout the installation.
- Make sure that controller and junction boxes are sealed to avoid insect intruders and moisture damage.

Box 9.3 Crimping tools and crimp-connectors

Crimping tools are special types of pliers used to ensure that electrical connections have good contacts. They securely attach bootlace-, ring- and spade-type connectors to the end of cables by pinching (crimping) the metal connector collar tightly around the end of the wire. Bootlace-, ring- or spade-type connectors attach more securely to terminals than bare wire. Because good connections are so important in DC systems, crimping tools and a good supply of crimp-connectors are an important part of any solar electrician's tool box.

Figure 9.3 *Crimping tools*

Mounting Solar Modules

Choose the array mounting location – and the method of mounting – during planning stages and not at the last minute! Choose carefully where to mount your modules. Pole mounts or ground mounts (for larger arrays) are usually preferable to roof mounts. For one to two modules, a pole mount is often the best solution: pole mounts enable easy orientation of the module and can be assembled in a workshop on the ground.

Roof-mounting of solar modules is often a problem for a number of reasons:

- Roofs are very different from each other so no single type of mount is the right one.
- Roofs face different directions.
- Roofs are shaded at different times of the day or year.
- Roofs are easily damaged and can be a dangerous place to work.
- Module-sellers mostly offer modules, not solar-module mounts.
- Customers often do not think about mounting modules until the last minute.

When there are more than two modules, avoid roof-mounting unless the roof is the only suitable site.

Solar cell modules should be mounted in a location where they receive the maximum solar radiation and where they will not be shaded, overheated, or covered with dust. They should be located as close as possible to the batteries and the charge controller, and in a place that is safe from vandalism and theft.

Read the instructions provided with the module before installing the mount. Make sure you know how to connect the wires before climbing onto the roof – it is difficult to read instructions when clinging to a ladder or roof!

Handling solar cell modules

Most solar cell modules are made with glass that is strong enough to withstand high winds or the impact of hailstones. However, like any other pieces of glass, modules can be broken by stones and they may shatter if they are dropped. Once broken, modules are usually impossible to repair. Even if only one cell is broken, the module will be ruined.

Thus, care should be taken when transporting modules. The back of modules especially should be protected during travel or work. Hard, sharp objects (such as screwdrivers) striking the back of the module can break a cell from behind. Avoid warping/twisting modules as this is likely to break cells.

If holes must be drilled in the frame for mounting purposes, take care not to punch through the frame into a cell (note that drilling holes in module frames can void warranties). Use a piece of wood behind the frame to prevent such an accident. Better still, use the holes that were drilled in the frame at the factory or do all drilling in a workshop with a clamp.

Figure 9.4 *Working on roofs can be hazardous; the workers in this photo are taking unnecessary risks. Use a safety harness if there is a risk of falls, and protective headgear when manoeuvring large modules and mounting structures into place (see section on safe working practices, pages 144–146).*

Choosing the mounting site

When selecting a mounting site, always look for an unshaded position that receives full sunlight throughout the day. Take the following points into consideration when deciding where to put your array.

Modules should be located:

- Off the ground, if possible, so that they are out of the dust and out of the way of humans and animals. Do not locate modules near chimneys or kitchens as they could become covered with soot and smoke.
- Where they will not be shaded between 9.00am and 5.00pm every day of the year. Check the position of the sun during different times of the day (and during

Figure 9.5 *Module mounting site*

different seasons) to determine whether shadows from trees, TV aerials or other objects will cross the intended module location. Note that even if one cell is shaded, the output of a crystalline module will fall considerably. The 'Solar Pathfinder' is a good tool for figuring out how shade from nearby obstructions will vary daily and seasonally (see Chapter 12 for information).

- Where they will not get too hot. If installed on a metal roof, module mounts should be at least 10cm (4 inches) above the metal surface, with enough space for air circulation behind the module. If possible, the module should be located where it will be cooled by the wind.
- In a secure location. Do not place modules where they might be stolen. Do not locate them where they might be vandalized or hit by stones.
- As near as possible to the batteries and controller. If the modules are located too far from the controller and batteries, there is a greater possibility of voltage drop and power loss so larger diameter and/or several cables will need to be used.

Method of mounting

Once the site has been chosen, it is necessary to decide which type of mount to use. Some solar electric suppliers and many installers provide mounts (or plans for mounts) and will be able to help decide which mount is the best for a given installation.

Most modules are mounted in a fixed position, and there are several options: on a pole; on the roof; or on the ground. Fixed mounts must be rigid, flat and well ventilated. They must also be strong enough to withstand the greatest winds anticipated for the location without bending or breaking. Tracking mounts are another option.

Figure 9.6 *Array mount designs*

- 'Pole mounts' are a good method to ensure modules are mounted at the proper orientation and location. They are good for small arrays (up to four modules). Pole mounts place the modules well off the ground in secure highly visible places. They require metal pipes (steel poles), strong frames and, usually, a foundation. Note that small pole mounts are easily converted into rotatable trackers for single module systems.
- 'Rooftop mounts' use racks or brackets to fix the array to the roof structure. Their advantage is that they are safe and secure, although it may be difficult to clean modules mounted high on the roof. Brackets may be constructed so that modules can be pivoted downward for easy cleaning or so that their angle can be adjusted seasonally. Never mount modules directly on to the roof without leaving space underneath them. Never make roof mounts the highest point on the roof, as they can attract lightning.
- 'Ground mounts' are used for arrays of four or more modules (e.g. for water pumps, refrigeration or large home systems with no other suitable place to locate the modules). They secure modules to racks fixed in concrete foundations and may be fenced off to protect the array from animals and curious people. Under normal circumstances, one- or two-module systems are not ground mounted. Make sure you have the correct compass positions when aligning the array. Fixed mounts should be aligned in a north–south direction. Also, use an inclinometer to make sure the array is at the proper tilt. A rule of thumb is to mount modules facing the Equator at an angle that equals the site's latitude plus 10°. In Zimbabwe, for example, modules should be mounted facing north and inclined at 25 to 30°. Check with local solar dealers to find what tilt and direction fixed modules are commonly mounted at in your area. Make sure that rain water can easily run off the modules (i.e. as a minimum they must have at least a 10° tilt). Never mount modules flat.
- 'Tracking mounts' can automatically track the sun throughout the day. However, they are generally not economical unless you have more than four modules. They add extra cost and complications to the system, and may increase system management work (or likelihood of breakdown). In Equatorial regions, a cheap, manually operated pole-type tracking mount can significantly increase module output. Manual mounts can be rotated two times each day to gain up to 25 per cent extra charge or seasonally to follow the sun as its orientation changes from north to south. Note that if the array is not rotated, power output will be considerably reduced!

Construction of mounts

Module-mounting structures can be purchased factory-made, or can be made to specifications by metal workshops, or can be constructed by installers. On large systems commercially manufactured mounts should be used – module suppliers will be able to advise. However, for systems of only a few modules self-made mounts or ones made in local workshops can be a good option.

When constructing all types of mounts (both fixed and tracking) corrosion-resistant, weatherproof parts should be used. Stainless steel and anodized aluminium angle irons are commonly used to make mounts. If the steel used to make the frame is not weatherproof, then it should be coated with a layer of red oxide paint to prevent corrosion and then painted. Remember, the mount should have a 20-year life. (For additional information about large array mounts, see Chapter 12). If wood is used to build mounts, it should be coated with a suitable varnish. Nuts, bolts and washers should already be weatherproof and corrosion-resistant.

When mounting modules on the roof, use ladders and walk boards to protect roofs from the weight of installers. Minimize the number of people on the roof. Bolt the mount securely to roof timbers or to the 'facia board' at the peak of the roof. Avoid tears in tin-roof surfaces that cause rain leakage (and mend any damage accidentally caused).

When installing an array with several modules, if possible first fix the modules to the frame on the ground and then attach the frame to the fixtures on the roof or pole.

Wiring the modules

Unless the batteries are being charged to power equipment during installation, do not attach the module leads to the battery or controller until the final connection sequence (see also the Final Connections section below).

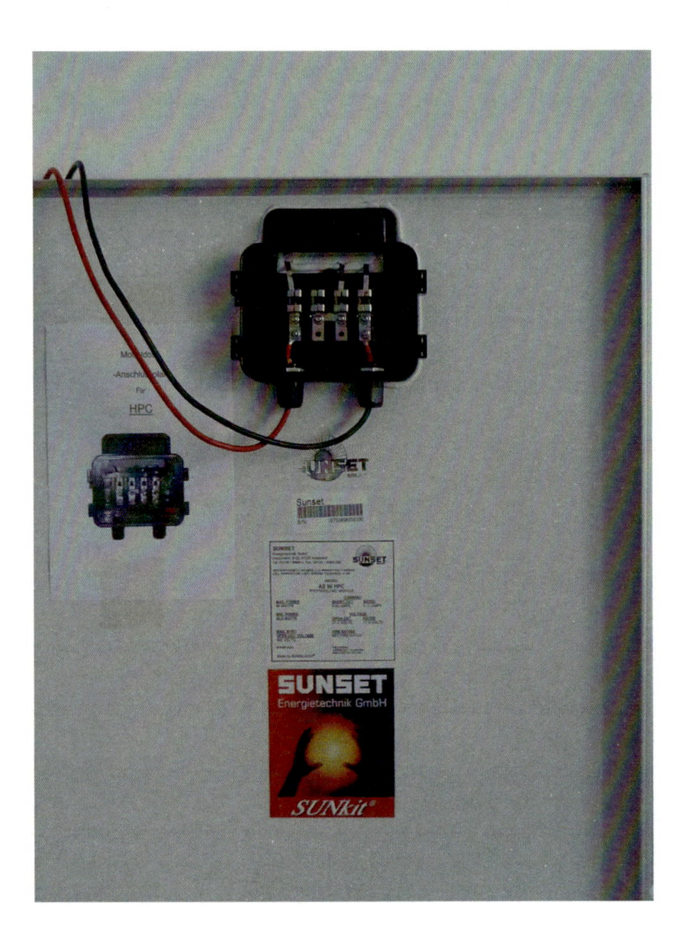

Figure 9.7 *Back of module showing leads from junction box*

- Remember, when wiring modules you are doing it for 20 years: take the time to do it right.
- Use correctly sized cable (see Chapter 7).
- When earthing is required earth the entire array by connecting the frame of each module to an earthing cable and connect this cable to the central earthing terminal (for more about earthing see Chapter 7).
- Wire module junction boxes carefully. Use crimp-connectors to make sure that the connections are good. Make sure that the junction boxes are well-sealed to prevent corrosion and insect intrusion.

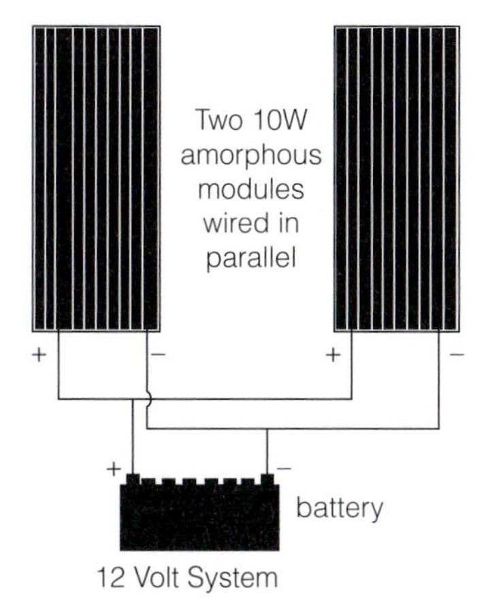

If there is more than one module in the array, then the modules must be wired in a configuration that matches the system voltage (The principles of series and parallel circuits are discussed in Appendix 2.) Two questions should be answered before attempting to wire modules together:

- How many modules should be wired in series?
- How many modules should be wired in parallel?

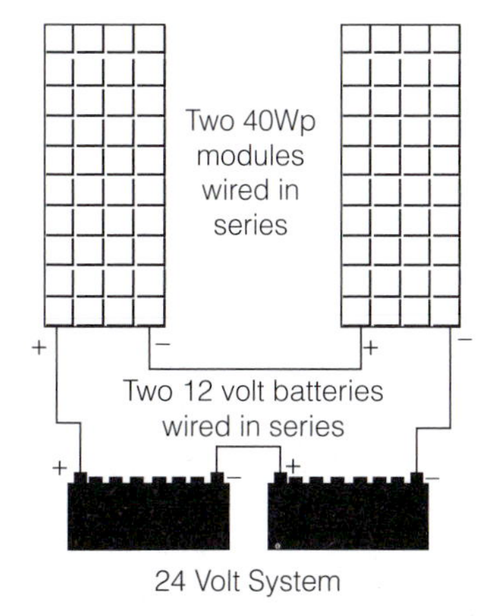

Figure 9.8 *Modules charging batteries in parallel and series*

Most small systems are 12 volt systems. Likewise, most commercially available off-grid modules (36-cell monocrystalline or polycrystalline modules) are suited to the charging characteristics of 12V batteries. Such modules are always wired in parallel in 12V systems. Figure 9.8 shows two modules wired in parallel to charge a 12V battery.

If the system voltage is 24 volts, then two modules (36-cell crystalline modules) must be wired in series to charge two 12V batteries in series. Figure 9.8 also shows two modules in series charging two batteries wired in series.

Battery and Controller Installation

The battery and controllers should be installed in an appropriate room or building. If there are only one or two batteries, they can be encased in a suitable box and placed in a suitable safe location (see below). If the system has three or more batteries, a separate room, closet or shed should by used to house batteries and/or the charge controller.

Choosing the battery location

Batteries should be located in a cool, vented room (see also Chapter 4). Here are some general guidelines for locating and installing the batteries, but for

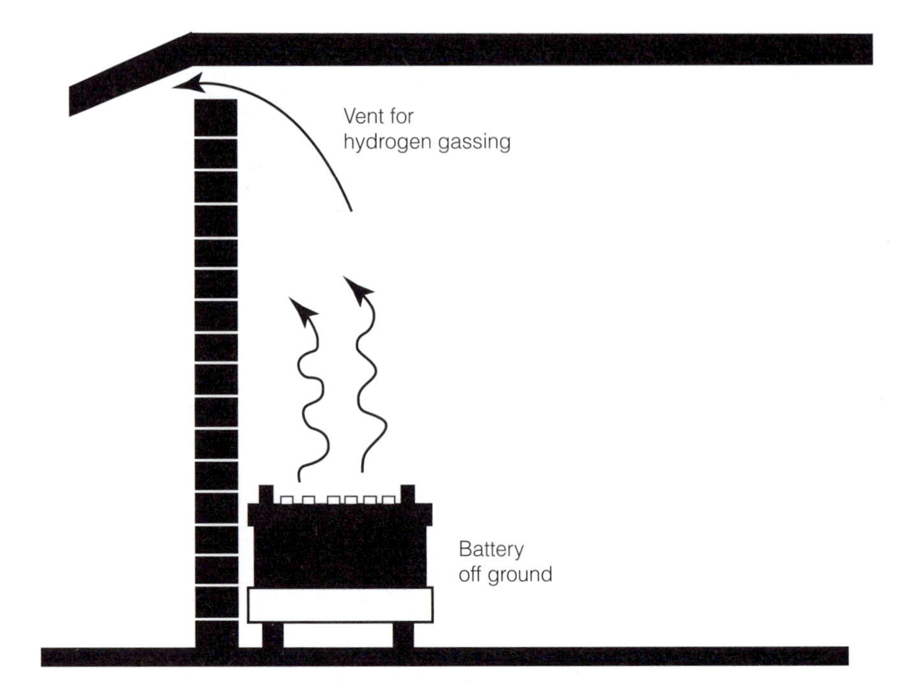

Figure 9.9 *The battery should be located in a place that is well-ventilated – including roof spaces – to make sure that hydrogen gas cannot build up between ceilings and roofs*

large battery banks (two or more units) the battery installation manual should be referred to.

- Nearness to array: batteries should be located as close as possible to the array to reduce voltage drop. The cable is generally sized large enough to carry the charge current from the module with 2 per cent or less voltage drop.
- Ventilation: the battery room must have some sort of opening for air to enter and leave. Batteries emit explosive gases when charging and this must be allowed to escape. Place a 'No Smoking or Naked Flames' sign in the room where the batteries are located.
- Accessibility: batteries should be accessible for easy state-of-charge measurement and cleaning, but only to authorized persons.
- Temperature: batteries should be located in a cool place. If battery temperature gets above 40°C (104°F), its lifetime and performance will be reduced. Never place a battery where it will be exposed to the sun.
- Battery boxes: batteries should not be placed directly on the floor as moisture or accidental puddles can increase their self-discharge rates. They should be kept in a vented box to prevent children and animals from injuring themselves and to prevent accidental short circuits (see Figure 9.10).
- Security and safety: locate batteries where they are secure and not likely to be stolen. A closet or room where children and animals cannot tamper with it might be suitable for a battery box containing two batteries. Battery boxes or rooms should be locked, but always ensuring that the key is nearby for quick access.

Figure 9.10 *Battery box, rack and shed*

Transporting and filling batteries

Lead-acid batteries are often supplied 'dry' for solar electric applications. This means that when sold, sulphuric acid electrolyte has not yet been added to their cells. Acid is supplied separately in plastic containers that can be safely sealed for transport to remote areas.

Because batteries are likely to tip and spill during transport on rough roads, if possible transport them dry and carry acid separately in sealed containers.

Also carry a large quantity of water to deal with possible spillages en route. If charged batteries are transported, place and cover them carefully so that they cannot tip over. Make sure their terminals are insulated and cannot come into contact with metal during the trip.

After arrival at the site, acid should be carefully poured into the batteries until they are almost full, either by using an acid pump (battery suppliers can provide one) or by using a funnel (always follow manufacturer's instructions if they are provided). Wait ten minutes for the acid to settle and then top them up. Afterwards, wipe the top of the batteries and thoroughly rinse the funnel, plastic tubes, clothes and hands that have come into contact with acid.

- Keep plenty of water around when filling batteries so that any spills can be quickly rinsed. Sulphuric acid is dangerous so take care when pouring it.
- Wear old clothes, plastic or rubber gloves and goggles when filling batteries. Fill them outside if possible.
- Once the batteries have been filled, they can produce a very high current. Keep tools and cables clear of the terminals and posts. Accidental short circuits are both expensive and dangerous.
- Remember to ensure that batteries are commissioned and fully charged by the array before starting to use them regularly. Battery manuals should provide full instructions.

Battery wiring

As stated previously, the cable run from the array to the controller and batteries should be as short as possible. If the modules are on the roof, run the cable through the roof space down an inside wall. If the modules are on a separate

Figure 9.11 *Battery terminal connections*

pole or ground mount, protect the underground cable run with a conduit and mark the location of the cable. Wiring codes will specify how underground cables are to be laid.

If there is only one 12V battery, attach the cables to the battery terminals (while the terminals are not attached to the battery), tighten the screws and secure them. It is always better to use ring- or spade-type connections on terminals than to simply wind the wire around the terminal screws (Figure 9.11). Unless the battery is being used to power tools and lights, leave one terminal disconnected from the battery until the final connection sequence. Coat the outside of the terminals with a thin layer of petroleum jelly. Check the electrolyte level in each cell. Check the state of charge of each cell to make sure that no cell is bad.

If there is more than one 12V battery, make sure that they are arranged properly in series or parallel. Make a drawing first, check that it is correct and follow it. Twelve volt batteries in 12V systems are arranged in parallel. If there are two 12V batteries in a 24V system, they should be arranged in series. If there are four 12V batteries in a 24V system, two should be in series and two should be in parallel (Figure 9.12). Connecting numbers of batteries up can be

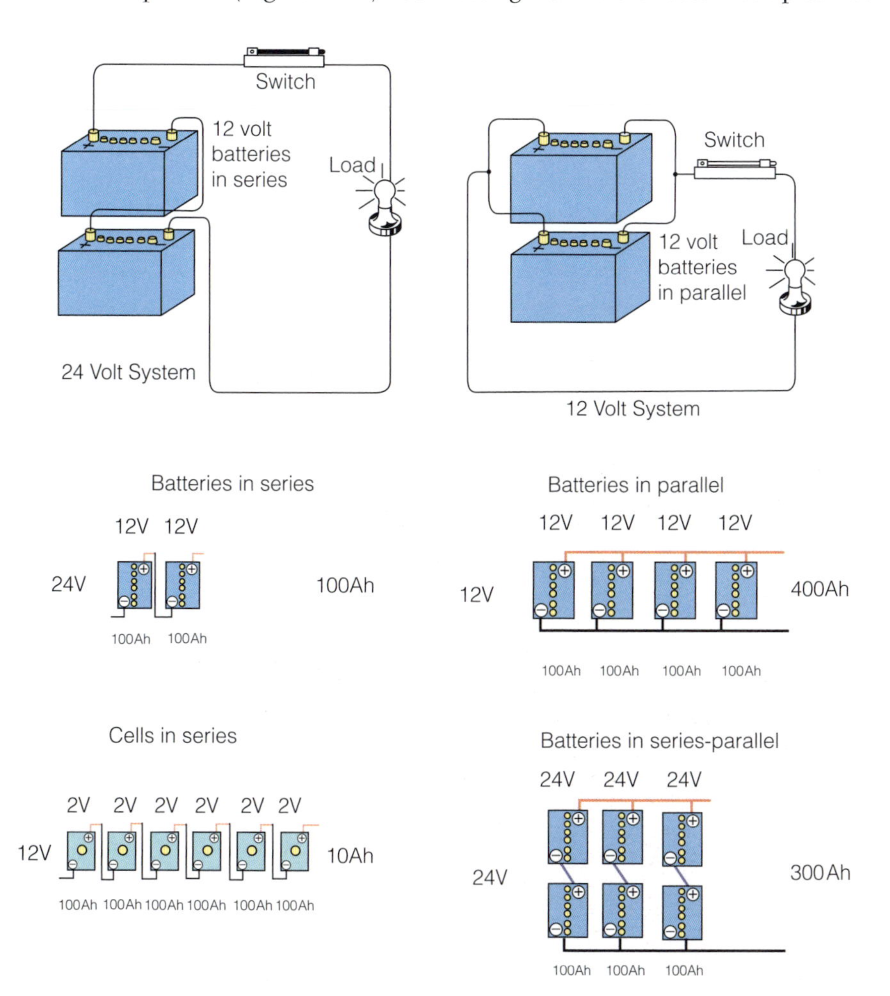

Figure 9.12 *Parallel and series battery configurations (12 and 24V)*

extremely hazardous and needs to be done by a properly trained technician. See Chapter 11 for large system battery configurations with 2V and 6V batteries.

Mounting the charge controller

The charge controller should be mounted in a location where it can be seen, as near to the battery and module as possible. Mount the controller on a sub-board or piece of wood – not the wall. It is easier to fix a board on the wall than a small controller device. Make sure you have the proper fastening tools before going to the site: fixing charge controllers and sub-boards to walls without proper screws or bolts is messy!

It is best to attach the charge controller, as well as the main switch, fuses and inverter, to a wooden sub-board. If the controller has none, a main switch and/or fuses can be fixed on the sub-board next to the controller (see Figure 9.13). All connections should be made in appropriate accessories or junction boxes.

Do not mount charge controllers (or inverters, or switches/disconnects) directly above the batteries. The acid mist from the batteries can corrode the charge controller circuits. Also, any electrical equipment containing switches/disconnects can spark – and the hydrogen gas given off by batteries is flammable.

Follow the manufacturer's instructions when installing charge controllers, as each type has its own procedures. Not following instructions can result in damage to the controller or malfunctions.

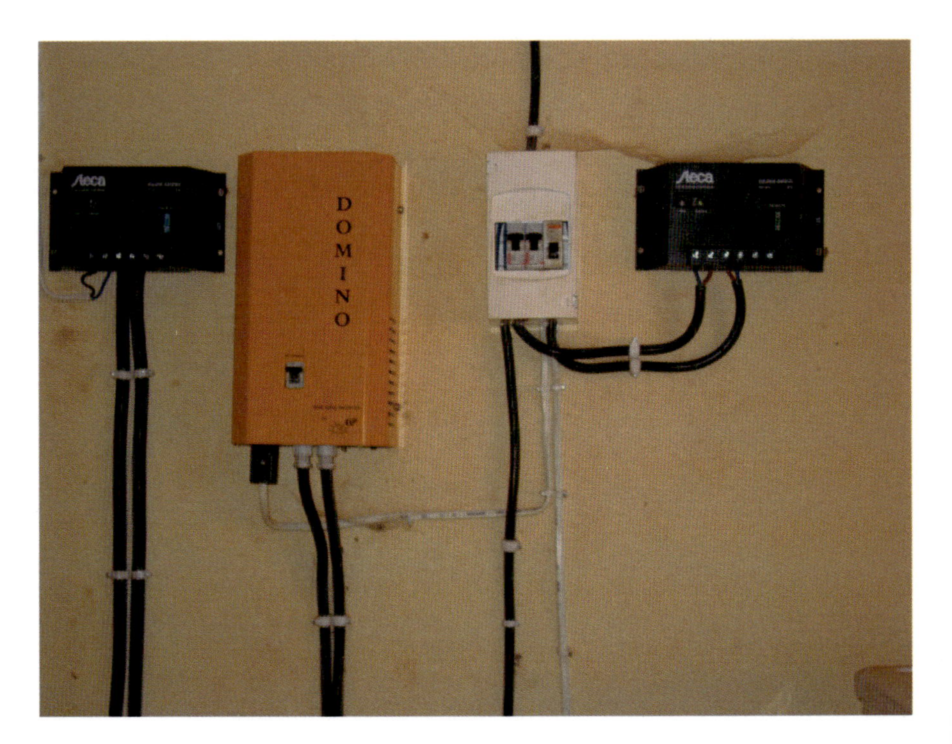

Figure 9.13 *Charge controllers, main switch and monitor on sub-board*

Testing and Final Connections

The guidelines below outline how to safely commission small systems:

- An electrical technician who is properly trained and experienced in solar electric systems should perform the tests and make the final connections. This electrician should conduct commissioning tests of the array, the junction boxes, the support structure, the solar modules, the control unit, the distribution wiring and the battery.
- In all cases, follow the manufacturer's installation instructions for modules, controllers and inverters, and comply with electrical codes.
- If the installation is for an institution or client, there should be an official handover to the operator/client with some type of receipt/record of transaction to mark the occasion. Completion certificates (often required by national codes) are a good idea. They should give details of the system and the results of all pre-commissioning tests carried out. A copy should be kept by the installer for future reference.

Wiring tests for small solar PV systems

Before making final connections, carry out the following electrical tests on the system. These electrical pre-checks will ensure safety, efficiency and proper operation. No electrical cables or system should be energized before it has been tested according to code requirements using the correct instruments – most codes will also specify the instruments to be used. The following is a summary of what is normally required:

- Array: check the module/array's open circuit voltage. Verify correct polarity/colour-coding of the cables coming into the building (with a multimeter set on the DC volt scale), and verify its short-circuit current (with the multimeter set on the 0–10 DC amp scale).
- Array-battery voltage drop: check the voltage drop between the module(s) and the battery using the procedure described in Chapter 7. Full sun and a substantial current flowing into the batteries are required for this test. A voltage drop of more than 2 per cent between the array and battery is unacceptable (some types of charge controllers reduce voltage so take this into consideration).
- Battery: check the battery voltage, DC polarity and fuse with a multimeter.
- Load circuit continuity: check all load/distribution circuits for continuity and check the switches to make sure they are working. A multimeter with a continuity buzzer makes checking switches easier.
- Test insulation resistance of the entire load/distribution circuits, with all switches on and all loads disconnected. This is to ensure the integrity of the insulation of cables and accessories. This test should be done with an insulation tester of the same type that is used on mains electricity circuits. It can be hazardous and damage equipment (tests are carried out at 500V DC) so it needs to be carried out by a competent person. In many places, and particularly for small SHS DC-only systems, insulation tests are carried out using multimeters; the problem with this is that a multimeter will only

identify very clear short circuits, not damaged cables, connectors or other accessories – all of which are fire hazards. In 110/230V AC systems insulation resistance tests will also identify shock hazards. Insulation resistance measurements should be greater than 500,000 ohms or whatever is specified in the local electrical code. Faulty insulation can cause fires.

- Outlet and socket polarity: the DC polarity of all lighting outlet and socket outlets should be verified (though this needs to be done when the system has been energized – see below).
- Earthing: the integrity and continuity of any earthing conductors should be checked.

Final connections

After conducting all necessary wiring tests and a comprehensive visual inspection of the installation the final connection sequence can be carried out. The order presented below is only a suggested sequence. Note that some charge controllers require a different connection order. Also, in some situations, the preferred procedure might be to connect up the modules to the batteries first (via the charge controller), then to connect the inverter so that AC power is available on-site while work is being carried out, and to do the main DC or AC distribution wiring later. Above all, have a plan and read all manuals before making final connections!

1 Connect the wires from the battery to the charge controller

Read the charge controller manual and follow the instructions. The main battery fuse should be taken out. Make sure that the wires are securely connected to the battery terminals and that the inside surface of the battery terminals is clean. Make sure the battery posts are clean. The negative battery cable should be connected to the negative (–) battery terminal of the charge controller and the positive wire should be connected to the positive (+) one. When this has been done replace the main battery fuse and measure the voltage across the charge controller battery terminals – it should be the same as the battery voltage.

2 Connect load and check load polarity

With the battery connected, power is available to check the wiring of lamps and sockets in the load. When the functionality of switches and polarity at lamp outlets has been checked then connect the light fittings. Now check that all DC socket outlets have a voltage and correct DC polarity.

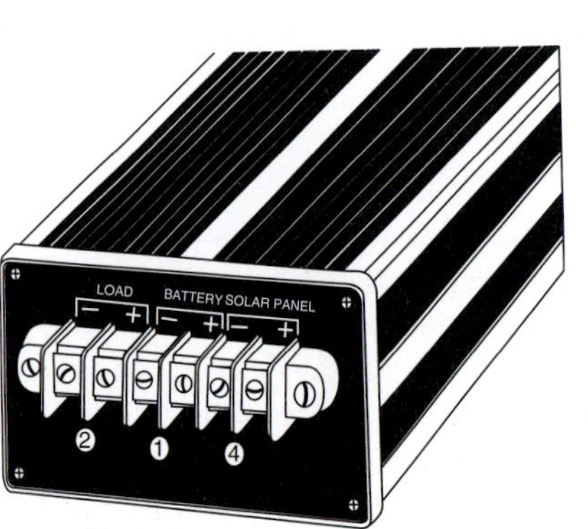

Figure 9.14 *Connection terminals for modules, battery and loads on charge controller*

3 Connect the fluorescent lamps and install the DC globes and LED lamps

Now that the polarity has been checked, the tube lights can be connected and put into service. Turn off the power to the distribution circuits before doing this. When connecting fluorescent lamp leads, check the wire coding for polarity (read the label for details). At this point, the DC globe and LED lamps can also be inserted into lamp holders. After all the lamps have been connected switch each lamp on, one at a time, to see if they are working properly and to check if the switch is properly installed.

4 Connect the solar cell module(s)

First, turn all the lamps and sockets off using the main switches in the controller (if there are none, turn them off at the room switches). The two main cables (+ and –) coming from the PV array will then be either connected to the charge controller (in a small one- or two-module system) or to an isolator/disconnect which feeds into the charge controller. These cables will be live so special care needs to be taken when verifying polarity and checking the open circuit voltage (Voc) and short-circuit current (Isc) of the PV array – which should be done before connecting up to the charge controller.

Warning: this can be hazardous! Make sure that the battery is connected to the charge controller first and that the main battery fuse is in – check that there is a battery voltage at the battery terminals of the charge controller, and make a note of it. Then connect the cables directly to the solar terminal of the charge controller or, if there is a PV array isolator/disconnect, switch that on – the cables from the controller and the PV array should be connected to the isolator/disconnect while it is in the open/off position. If the sun is shining and there is a charge indicator LED on the controller it should become illuminated and the battery voltage should rise between 0.3 and 0.4V (depending on how much sunshine there is, though if the battery is full it may not accept a charge). If there is a problem, disconnect the PV array from the controller immediately, check all wiring again and refer to the charge-controller manual.

5 Final visual inspection

Check to see that all the lights and sockets are in working order (plug an appliance into each socket to test it). Using a voltmeter, check for voltage drop at the farthest point from the batteries. Check the LED readings on the charge controller.

The system wiring is complete, and it can now be safely commissioned and used. Remember, batteries should be fully charged before the system is utilized.

User Training

Chapter 10 discusses solar electric system maintenance. Certainly, the tasks involved in managing a small PV system are a lot easier than those involved in managing a generator or car. However, these tasks do need to be done and energy use does need to be managed. System success depends on the proper training of users to manage and maintain the system. Since PV-system installers often do not operate or maintain the systems they install they must train the operator to manage the system.

System management and maintenance routines

As is the case with a car or motorcycle, there is a set of routines that need to be followed when a small solar PV system is operated, serviced and maintained. As discussed in Chapter 10, routine tasks need to be completed on a daily, weekly or seasonal basis. As is the case with a car, someone needs to take responsibility for these tasks, and make sure they get done.

- Operation routine: users need to be aware of operation routines. This means regularly checking charge controller indicators and managing loads.
- Expected maintenance and service schedule: operators must know which components need regular attention (e.g. batteries) and which parts wear out and require replacement. Operators must know which spares are required (fuses, bulbs, distilled water) and where they are available. If there is one, the system manager should be given a copy of the service contract.
- Record keeping: in some systems, particularly for institutions, records should be kept regarding the age and condition of the batteries (they will have to be replaced), the place of purchase of system components and the electrical details of the system (i.e. circuit diagrams). The manager needs to be shown how to keep these records.
- What to do in case of breakdown: the system operator should know who to contact in the case of a problem that cannot be solved on-site. There should be some kind of regular contact (at least annually) between the operator and the sales agent or installer.

Load management

The system operator/manager must make sure that the energy from the system is utilized so that the daily system energy demand is in balance with the harvest of electrical energy from the sun. This means actively controlling the use of the electrical energy.

- Appliances should be used according to the plan for their use (see Worksheet 1). If a TV is supposed to be used for four hours according to the worksheet, then eight hours of daily use will inevitably drain the battery and cause the charge controller to disconnect the loads.
- The system manager should ensure that the batteries get a full charge at least once a week.
- The number and hours used of connected loads should not be allowed to creep upwards. If the system is installed in an institution, then the establishment of log-sheets may be useful to track energy use.
- During extended cloudy weather, the system manager should reduce energy use in line with the reduced electricity harvest.

10

Managing, Maintaining and Servicing Off-Grid PV Systems

This chapter explains how to take care of solar electric systems and how to fix them if they break down. It explains routine maintenance tasks involved in the care of the batteries, modules, wiring, inverters and charge controllers, and loads. The section on troubleshooting explains how to identify the causes of problems and how to solve them.

Managing Energy Flow in Off-Grid PV Systems

When off-grid solar electric systems fail, the most common cause is poor system management or poor design and installation. Too often off-grid PV system owners do not understand how much energy their system is collecting and how much energy their appliances are using. Consequently, they manage energy poorly and their batteries are continually in a low state of charge. Continually discharged batteries are a major cause of battery – and system – failure. The section below is intended to help you gain an understanding of energy management in PV systems.

Collecting and storing more energy

The following tips are common sense. If followed rigorously, they will help to ensure your system collects and stores energy effectively.

- Keep batteries well-charged. Well-charged batteries have longer life-spans and do not have to be replaced as frequently. Ensure the battery gets to a full state of charge at least once a week!
- Keep batteries clean. Clean dust and acid mist off the top of the casing. Keep terminals corrosion free. Corroded terminals cause voltage drop, loss of energy and eventual system failure.
- Keep modules clean. Dust-covered modules produce less electric charge. Regular cleaning of the module (i.e. every two weeks) in dusty locations will pay off in terms of extra energy collected. Always mount modules where they can be easily reached for cleaning.
- Ensure proper connections and cable choice. Choose correct cable sizes during planning to avoid voltage drop. Make sure that connections are tight, especially at the controller, module, inverter and battery terminals. Poor contacts cause voltage drop and loss of energy, and are also a fire hazard.

Box 10.1 Tracking and adjusting solar arrays: getting the most energy from solar modules

'Tracking' is changing the position of the solar array to face the sun as it moves across the sky from morning to evening or as it changes its path through the sky according to season. This management practice helps to get the most power from the sun and is especially beneficial to those close to the Equator. Investment in automatically powered solar trackers can be useful for arrays of above 500Wp (see Chapter 12 for information about commercially available trackers). This section refers to manually operated tracking mounts for small arrays or seasonally adjusted fixed mounts.

There are two types of adjustments that can help track the sun:

- Daily adjustments.
- Seasonal adjustments.

Rotatable pole mounts

For small systems located near the Equator (i.e. between 15°N and 15°S), users may want to consider using simple manually operated rotatable mounts. These mounts, which cost much less than the price of a module, can produce upwards of 20 per cent more solar charge. With this type of mount, the module is mounted on a bracket (at 25° near the Equator) attached to a steel pole (see Figure 10.1). The pole is rotated twice a day: first to face the morning sun's 10.00am position and, at 1.00pm, to face the afternoon sun's 4.00pm position.

Figure 10.1 *Using a rotatable tracking mount*

In Equatorial locations, rotatable pole mounts enable users to harvest more energy in different seasons. The pole mount can be turned to face the sun when it is in the south (i.e. October to March) and likewise when it is in the north (April to September).

However, be warned that if the mounts are not operated properly each day, the module(s) will collect considerably less energy!

Seasonally adjustable mounts

Those installing PV systems outside of the tropics may want to consider making their fixed north- or south-facing mount adjustable so that it collects optimum radiation depending on changes in the sun's path through the sky. This means increasing or decreasing the horizontal angle of the array so that the sun's rays strike the modules at an angle that is closer to perpendicular. Changing the angle of the mount four times per year yields about 10 per cent more output from your array over the course of the year in northern or southern latitudes. Importantly, it can help provide more energy in winter seasons when it is most needed (see Chapter 12 for more resources on adjustable tracking mounts).

As an example, Table 10.1 shows how an operator of a solar system at a site at a latitude of 45° north would adjust their mount to follow the path of the sun in the sky.

Table 10.1 *Example of angle adjustments of a fixed solar array at 45° north latitude*

Date to adjust array angle	New angle
Mid-February	45° facing south
Mid-April	22° facing south
Mid-August	45° facing south
Mid-October	65° facing south

Figure 10.2 *Seasonal angle adjustment of a fixed mount*

Active Management of Collected Solar Energy

Load management is the practice of making sure that energy demand balances the harvest of electrical energy from the sun. This means that use of electrical energy must be carefully monitored and controlled, especially during cloudy periods. Properly functioning off-grid PV systems are actively managed by their owners: when there is excess solar, this is utilized, and when available energy is constrained, energy use is reduced.

Understand systems' energy collection and use limitations

All off-grid electrical systems need to be managed; some will need very little management and some will need quite a lot. Controllers and monitoring devices will help manage a system. They are tools that are best used by an educated system manager.

Use the information in the system-planning section (Chapter 8) to understand energy flows in systems. System owners need to be aware which months have the lowest energy availability and take actions to reduce consumption during those periods. Meters or indicators on charge regulators should be checked regularly so that the state of charge of the batteries is known. Good managers keep weekly records of their battery state of charge!

Remember, someone must manage the energy use of any off-grid PV system. Managers of PV systems should decide which energy requirements are most important. When installing a system, make sure that there is an energy use plan in place and ensure people know why there is a need to conserve energy.

Use loads according to the system design and turn off appliances not in use

Systems are designed to produce a given amount of electricity. Appliances should be used according to the plan for their use (see Worksheet 1). If a TV is supposed to be used for four hours, then eight hours of daily use will inevitably drain the battery! Remember also, an 8W bulb left on accidentally for a few nights can be a major drain on a small system. Loads that are not in use should be turned off.

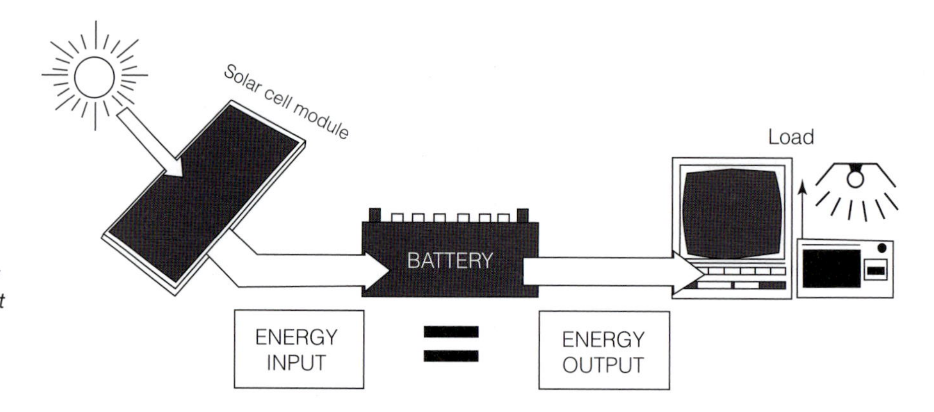

Figure 10.3 *The energy output to the loads must be balanced by the energy input from the array*

Figure 10.4 *Remember to turn off lamps and appliances when not in use. An educated system manager is far cheaper, and, in the long run, far more effective than any control device.*

'Phantom loads' or 'ghost loads' should be turned off or unplugged when they are not in use. Phantom loads are circuits in appliances that 'steal' power even when the appliance is switched off. They include inverters, power adaptors from laptops and cell-phones, remote-control sensors in TVs, or LED indicators on printers and other office equipment.

Reduce power use during cloudy weather

During cloudy months the array output is likely to be reduced by one-third or more. The load may therefore have to be reduced to protect the batteries. System managers need to consider which loads need to be reduced. If the television or computer is the most important load, then reduce use of the lights until the weather turns sunny again. If the lights are the most important loads, then reduce or cut-out use of other appliances.

If a system has only one battery, for example in a small solar home system, consider taking it for charging during especially cloudy periods, but make sure the charger is suitable for charging the battery (check with your battery supplier). In small one-battery systems, it is generally a good idea to give the battery, if it is an automotive wet cell – not a gel cell – a 'hard' charge from a grid or generator-powered battery-charger twice a year. It is much cheaper to charge a battery under stress than to replace a ruined one. In larger systems, with several batteries, this is hardly practicable so consider buying or renting a small generator battery-charger for your site during long cloudy periods (again, make sure the battery-charger is suitable for charging the type of batteries involved).

Remember to let the batteries reach a full state of charge every time they have been run down to a low state of charge.

Monitor and control load-demand increases

A common problem with off-grid PV systems is that there is a 'creeping' increase in load demand. Often, after a period of well-managed use, energy demand creeps upwards. People add appliances to the system without thinking

Box 10.2 System records and manuals

System records help those who are managing and maintaining solar PV systems. For example, troubleshooting electricians may need to see the electrical drawings of the system. Records of daily battery state of charge and module output can help analyse problems when they turn up and can help avoid more serious ones.

Keep all information about the system in a safe place, preferably under lock and key, where it can be referred to when necessary. Update it periodically. Most of the important information can be kept in one ledger or file. Large institutional systems work better when someone is given the job of maintaining the system and keeping records up-to-date.

Important PV system information includes:

- Circuit diagrams and maps showing the location of batteries, loads, wire runs, junction boxes and buried cables.
- The system completion, inspection and test certificate.
- System design worksheets, plans and specifications.
- Manuals, warranties and manufacturers' specifications for system components.
- The system log book. This contains records of battery state of charge and history, installation dates, repairs, equipment replacement and system maintenance.

and start to use them for longer periods. System managers must be vigilant to prevent or plan for this.

- Keep records of daily battery voltage and any other parameters that your system metering allows. Do this in a log book (see Figure 10.5).
- Plan for additions of appliances and loads! If you add appliances, increase the size of the solar array and battery bank according to the increase in the load.
- Talk to all energy users to make sure they understand the limitations of the system. Encourage them to understand how the system works.

Technical Solutions to Reduce Energy Use

Use efficient appliances and lamps

Fluorescent tube and LED lamps are always preferable to incandescent lamps. When buying appliances such as radios, televisions, computers, refrigerators or sewing machines, choose types that use less power but still meet your requirements (see Chapter 6).

It is usually more expensive to power inefficient appliances (that need a large array) than it is to spend the money on more efficient, but more expensive, appliances (that need a comparatively small array). For example, although laptop computers and super-efficient DC refrigerators are more expensive, they save enough solar PV power to justify their costs over a short period of time.

TECHNICIAN'S NAME:				
SITE DETAILS:				
Battery Voltage (when arriving)	VDC			
Time: 6 7 8 9 10 11 12 13 14 15 16 17 18	Sun condition	☐ bright	☐ cloudy	
Controller Battery indicator	☐ High	☐ Low	☐ Cut off	
Time: 6 7 8 9 10 11 12 13 14 15 16 17 18	Sun condition	☐ bright	☐ cloudy	
Battery Voltage (with solar disconnected and loads OFF)	VDC			
Time: 6 7 8 9 10 11 12 13 14 15 16 17 18	Sun condition	☐ bright	☐ cloudy	
Battery Voltage (with solar disconnected and loads ON)	VDC			
Time: 6 7 8 9 10 11 12 13 14 15 16 17 18	Sun condition	☐ bright	☐ cloudy	
Battery Acid Level	☐ OK	☐ Low		
Battery Terminals	☐ Clean	☐ Corroded		
Top of Battery	☐ Clean	☐ Dirty		
Solar Module(s)	☐ Clean	☐ Dirty		
Solar Module voltage (Voc)	VDC			
Solar Module Current (Isc)	A			
Time: 6 7 8 9 10 11 12 13 14 15 16 17 18	Sun condition	☐ bright	☐ cloudy	
Fluorescent Lights (number)	#_____ OK	#____ Black		
DC/DC Converter	☐ OK	☐ Faulty		
Fuse(s)	☐ OK	☐ Blown		
Inverter	☐ OK	☐ Faulty		
Switches	#_____ OK	#____ Black		
Sockets	☐ OK	☐ Faulty		
Battery Voltage (when leaving)	VDC			
Time: 6 7 8 9 10 11 12 13 14 15 16 17 18	Sun condition	☐ bright	☐ cloudy	
Controller Battery indicator	☐ High	☐ Low	☐ Cut off	
Time: 6 7 8 9 10 11 12 13 14 15 16 17 18	Sun condition	☐ bright	☐ cloudy	

Work performed and Comments.

Please indicate if any part of the system has been tampered with.

____

____

____

____

____

____

NEXT SERVICE IS DUE ON	

Figure 10.5 *Sample system log book*

Put reflectors on lamps and paint walls with a light colour

In situations (such as schools, clinics or workshops) where solar power is primarily used for lighting, place reflectors on lamps used in work areas. Reflectors direct up to 40 per cent more light from the fixture to desks or work-benches than lamps without reflectors. White-coloured walls will make any room much brighter.

Use timer switches on lamps and loads

Timer switches turn lights and appliances on and off automatically so that energy is not inadvertently wasted. There are two types of timer switches:

- Controller-based timers. Located in the control system, this type of timer turns major circuits on and off at times set by the system manager. This is particularly useful in schools or workshops. Such timer switches can be provided with a manual override so that the system manager can occasionally leave lights or loads on for a few extra hours.
- Appliance-based timers. This type of timer is simply a switch with a timing device inside it that automatically turns the light off after a preset amount of time (e.g. five minutes). Such switches are commonly used in European bathrooms and hallways to save energy lost from lights carelessly left on.

Routine Maintenance Tasks

A properly installed solar electric system requires relatively little maintenance. In fact, the work involved in maintaining a solar electric system is much less than that needed to maintain a diesel- or petrol-powered generator. The best maintenance practice is to make regular inspections of the equipment (especially batteries and modules), to make sure things are kept clean and to make sure all electrical contacts are tight.

The following section describes most of the tasks that need to be done when maintaining a system. If the suggested procedures below are followed, a system

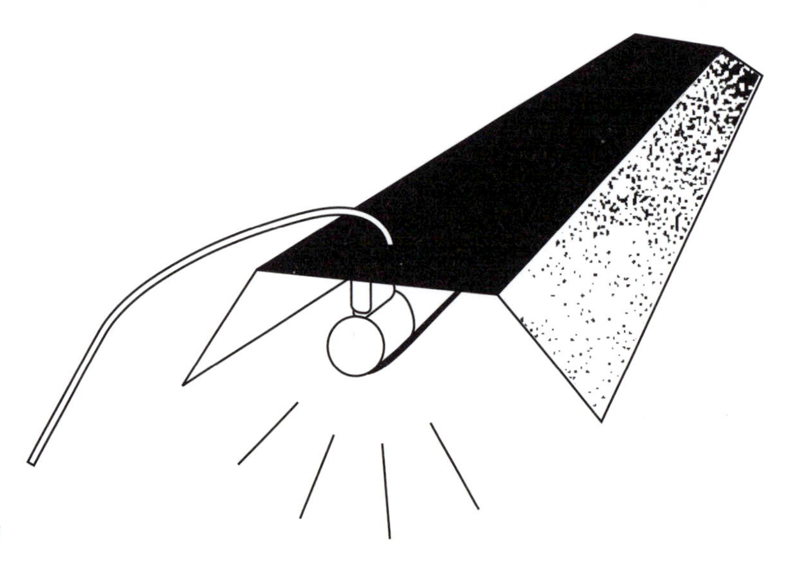

Figure 10.6 *Reflector fitting*

should work well. Perform an annual system check to look for problems not covered below, such as tree-growth or birds' nests shading the modules, insects in junction boxes, garden plots above buried wires and other unexpected problems.

The items mentioned in Table 10.2 should be kept on hand in addition to the tools mentioned in Table 9.1.

Analysis of system records

During the routine maintenance and servicing of any system, look through the operation records of the system (if they are kept). System managers should check the following on a monthly basis:

- Daily battery state-of-charge log. This will indicate whether the battery has been in a low or full state of charge, and whether this continued for a long time.
- Records of any recent repairs or problems with the system.
- Some charge regulators and amp-hour meters store detailed records of battery state of charge, daily energy use and solar charge (a few of these enable the information to be downloaded on to memory sticks or laptops). Always review this data to see how the system has been performing, and how much energy has been collected and utilized.

Battery maintenance

Batteries are the heart of any off-grid PV system. They should be kept in a high state of charge (above 70 per cent SoC at the very least). For long life, they should be allowed to reach full charge at least once a week. After deep-discharges, they should be allowed to fully recover their charge. They should be kept clean. Flooded batteries should have their electrolyte level checked regularly.

Table 10.2 *Useful tools and materials for maintenance of solar electric systems*

Tool	Purpose of tool
Insulation tape	Making emergency repairs on wires
Hydrometer	Measuring battery and cell state of charge
Multimeter	Measuring state of charge, checking wiring
Adjustable spanners	Tightening loose connections
Distilled water	Replenishing battery electrolyte
Petroleum jelly	Protecting battery terminals
Baking soda	Neutralizing spilled battery acid
Spare switches	Replacing broken switches
Spare tubes and globes	Replacing burned out lamps
Spare fuses (of the proper size)	Replacing blown fuses
Extra screws and wires	Replacing stripped or worn screws and wires
Connector strips	Replacing worn or broken strips
Sandpaper	Cleaning corroded battery terminals
Spare battery terminals	Replacing corroded or damaged units

Wear eye-protection and rubber gloves when maintaining batteries. Be extra careful when coming into contact with battery acid. Be aware of the possibility of short circuits when maintaining and servicing battery terminals. Never smoke or carry a flame around batteries. Always wash your hands immediately after handling batteries – traces of acid on your hands will burn holes in clothes and irritate your skin.

Check battery state of charge (at least twice a month or as required by the system log)

For information on checking battery state of charge see Chapter 4. If the battery is in a low state of charge reduce use of the load and allow the battery to be fully charged by the array (or have it charged by alternative means). Check battery state of charge at the same time every day. A good time to do this is when the sun is low in the sky, and when no loads are in use. For large systems (e.g. institutions), keep records of battery state of charge and performance in a log book. This allows users to judge more easily whether a battery needs replacement or whether it is failing.

Flooded batteries require more intensive maintenance than AGM or gel batteries, but all batteries require regular maintenance.

Clean batteries (check once a month to see if cleaning is required)

If you are cleaning a single battery, carry it outside to avoid spilling acid. Keep plenty of water nearby to rinse spills. If you are cleaning a bank of batteries, make sure the battery enclosure is clean and dry, in addition to the batteries. Use a clean rag and sandpaper for battery cleaning tasks.

1. First, switch off or disconnect the solar charge.
2. Remove the battery fuse and disconnect the battery from the leads, and remove the battery terminals from the posts (see Figure 9.11).
3. Clean the top and outside of the battery with a cloth and water (do not allow water to enter the cells).
4. With sandpaper, clean battery terminals and posts until they are shiny. If the terminals are corroded (i.e. if they are covered with white powder), clean them carefully using sandpaper and a solution of baking soda and water. If the terminal has been badly corroded replace it.
5. Put back the clean terminals and tighten the bolts. Apply petroleum jelly or grease to protect the outside of connected terminals.
6. And don't forget to wash your hands afterwards.

Check and top-up electrolyte level if required (once a month)

This is only required with flooded batteries (see Chapter 4). When topping up batteries use distilled water only, which can be obtained from petrol stations (do not use tap or river water). Wear protective goggles while checking and topping up electrolyte levels.

1 Remove the caps of each cell one at a time and check the level of the electrolyte. Acid should be within 2cm (¾ inch) of the top of the battery (or at the indicated level). If you can look inside the batteries, check the plates to see their condition. Make sure the acid is well above the level of the plates.
2 If the electrolyte level is below the required level, add distilled water until the electrolyte is at the indicated level, about 2cm (¾ inch) below the cap opening in modified SLI batteries.

Module maintenance

Since modules have no moving parts, they require minimum maintenance. Keeping the glass surface clean is the most important task. Also, be aware of shade from plants or trees that grow up around the array. Check occasionally for loose nuts and corrosion in the mounting hardware.

Clean dirt and dust build-ups on modules (once a month)

1 Solar modules must be kept clean to produce maximum power. Dust collecting on top of the module can greatly reduce electric output. During the dry seasons, inspect the module every two weeks for collected dust by running a finger along the top. Also, look for bird-droppings, leaves, streaks or signs of damage.
2 Clean modules with water and, if necessary, a mild soap (solvents should never be used). Wipe the glass with your hands, a sponge or a soft cloth (rough cloths or brushes will scratch the glass – do not use). Make sure all surfaces of the array are fully rinsed and streak-free.

Check array output and module/array connections (once a year)

1 Check the output of the array at noon on a sunny day. Measure the open circuit voltage (Voc) and the short-circuit current (Isc). Compare these with previous figures, and note any change.
2 Inspect the junction box on the back of each module to make sure that the wiring is tight. Make sure that wires have not been chewed or pulled out. Ensure that no insects or lizards are living in the junction boxes. Remove plant growth, spider webs and debris from the back of the array.

Wiring and charge controller

If wiring is installed properly, there should be no wiring problems for the life of the system. However, it is useful to check the wiring of a system at least once a year, especially in places where it might be chewed, tampered with or accidentally pulled.

Inspect wiring, fuses, indicator lamps and switches (once a year)

1 Inspect system wire runs for breaks, cracks in the insulation or places where it has been chewed. This is especially important for old or exposed wire.
2 Inspect junction boxes to make sure they have not become homes for insects. Make sure they are still watertight.
3 Check switches to make sure they are operating properly.

4 Check fuses to find if any have blown. If so, find the cause and repair. Replace a blown fuse with a new one of the same size.
5 Check the indicator lamps on the charge controller. The solar charge indicator should come on when the sun is up. If it is not on, check to see if the batteries are being charged. Check whether the other LED indicator lamps are working.
6 Check the tightness of connections in the terminal and connector strips. Make sure no bare wires are visible.
7 Check earthing cables and rods to make sure they are still intact.

Lamps and loads

On a daily basis, operate the loads as efficiently as possible. Maintenance of loads includes turning lights and appliances off when not in use. As needed, do the following:

1 Clean lamps, reflectors and fixtures once every few months. Dust and dirt make lamps seem less bright.
2 Check for blackening tubes in fluorescent fixtures. If tubes blacken at one end it is an indication that they are approaching the end of their lives and that their output is reduced. Replace blackened or blinking tubes.

Troubleshooting

Troubleshooting means fixing problems as they occur. Although the equipment in properly installed systems is unlikely to fail, problems sometimes occur. This section explains how to tackle problems in solar electric systems when they do occur.

First, don't panic. Most problems have very simple causes and can be discovered simply by checking in a few key places. The battery is a likely source of problems in small solar electric systems and it is likely to give you clues on the cause of the problem.

Always carry a multimeter when troubleshooting, as you can use it to quickly measure the battery's state of charge, check for broken wires and shorts, check the output of the module and measure voltage drops. Fuses can be easily checked with a multimeter: an intact fuse has a low resistance, less than 1 ohm – a blown fuse has a high resistance (mega ohms). Learn where to buy fuses and electrical equipment. The section below provides basic questions you should first ask about the system to identify the source of the problem. This is followed by a detailed table that should help you identify specific problems.

Check for basic problems first

- Read the log book. This will have vital information about recent battery state of charge and servicing.
- What was the weather like for the weeks before the problem? Has the weather been cloudy? Is it likely that the load has been using more energy than the solar modules produce? If this is the case then the problem may be a management issue and not a failure of any part of the system.

- Is the system new? Do the owners know how to use and maintain it properly? If the system is only a few weeks old or less, then the problem may be due to the failure of one of the parts (due to faults in the components) or improper installation. On the other hand, if the owners do not know how to use the system, you should question them carefully about how they manage the system.
- What is the type, condition and age of the battery? Can it still hold a charge? If the battery is corroded and looks like it has not been cleaned in months, then it may be the source of the problem. Similarly, if the system uses a five-year-old battery, there is reason to suspect that it has reached the end of its life. If, however, the battery is new, clean and well-charged, then you may have to look elsewhere.
- Are all the fuses and circuit breakers okay? Locate all the fuses in the system and see if any have blown. Check to see what caused the fuse to blow (e.g. overload, short circuit) before replacing it.
- Are all the wires connected securely? Are any corroded? Is there any place where a wire is likely to have broken?
- Are the modules dusty? Are they shaded? Is one missing or broken?
- Is the charge controller functioning? Is it delivering solar charge to the battery? Are all the indicator LEDs or meters working?

Detailed troubleshooting guide

If you cannot find the problem with your system after using the above basic check, then you may have to do a bit more exploring to find what is wrong. The following troubleshooting guide (Table 10.3) outlines other possible causes of system failure.

Table 10.3 *Troubleshooting guide*

Problem	Cause	How to Fix
1. Battery State of Charge is Low "Battery low" indicator LED on, low voltage disconnect turns OFF load, or battery state of charge is constantly below 11.5 volts	• There is no solar charge	See #2 (No Solar Charge) below
	• Battery acid low	Add distilled water to cells
	• Bad connection to control terminal	Check for broken wires or loose connection
	• Defective (bad) battery or cell	Check state of charge of each cell. If there is a significant difference between cells, replace or repair
	• Loose or corroded battery terminal	Clean and tighten battery terminals
	• Dusty modules	Clean modules
	• Blown fuse	See "blown fuse" section, below
	• Overuse of system	Leave appliances and lamps "OFF" for a week to allow recharging or recharge battery by other means
	• Battery will not accept charge	Find out age and history of battery. Check manufacturer's data. Replace all batteries if bank has exceeded expected cycle life.
	• Voltage drop between module and battery too high	Check voltage drop. Replace cable with proper diameter if required.

Table 10.3 *Troubleshooting guide* (Cont'd)

Problem	Cause	How to Fix
	• Defective controller	Check operation of controller with dealer. Replace or repair if necessary.
	• Inverter draining battery	Run system without inverter. Reduce AC load use.
2. No Solar charge Solar charge indicator does not light up during the day. There is no current in wires from array.	• Short circuit along wires to modules	Locate and repair short circuit
	• Loose connection in wires connecting battery to the control	Locate and repair loose connection
	• Blown "solar" fuse	See #5 (Blown fuse) below
	• Thick coating of soot or dust on module	Clean module with water and soft cloth
	• Broken module	Check for broken cells, broken glass, or poor connection inside module. Replace solar cell module.
3. Lamps do not work One or more lamps fails to come ON when connected. (Check for blown fuse first)	• Bad tube or globe	Replace with new tube or globe
	• Broken ballast inverter (fluorescent lamp).	Replace ballast inverter with new one
	• Bad connection	Locate broken or loose wire and repair
	• Switch is "OFF"	Turn switch "ON"
	• Tubes or globes have very short lifetimes	Check voltage of system: too low or too high? (Voltage is always lower when load is ON)
4. Appliances do not work One or more appliances fails to come ON when connected or operates poorly. (Check for blown fuse first)	• Bad connection	Locate broken or loose wire and repair
	• Switch is "OFF"	Turn switch "ON"
	• Defective/broken socket	Check socket. If defective or broken, replace. Check fuse in socket
	• Broken appliance	Try appliance where there is a good power supply. Replace or repair
	• Inverter not working (For AC Appliances)	Turn inverter "ON" Repair/replace inverter
	• Poor operation of appliance	Check for low voltage (DC) Check for output of inverter (AC)
5. Blown fuse The fuse is removed, fuse wire is melted.	• Short circuit along wire to solar cell module battery or load	Locate cause of short circuit, repair – then replace fuse.
	• Fuse sized too small	Use fuse 20% larger than combined power of loads
	• Lightning strike or power surge	Replace fuse

11

Basics of Large Off-Grid Systems

This chapter is about the special needs of 'larger' or 'specialized' PV systems for applications in off-grid areas. For skilled engineers or electricians, installing small PV systems is a straightforward process, as energy flows are small, the number of components are few and the design principles are simple. However, when designing and installing larger systems, other factors come into play. Arranging and sizing many modules and batteries requires experience, design skills and installation skills. Control systems are more sophisticated and they should also tell the user more about the status of the system. Different power sources such as diesel generators and wind generators may be incorporated in the system. Because of potentially dangerous voltages and currents, large systems must be designed to strict safety-standards codes. This chapter contains four annotated case studies of solar PV systems that may help designers plan systems.

From Smaller to Larger Systems

One of the most satisfying aspects of the single module 'solar home system' is that it is so simple. A 50–100Wp PV system will readily power a small household's DC lights, DC radio, DC mobile-phone charger and small DC TV with only the minimum of system management required. For more than a billion people around the world, simply having electricity for lights and communication would greatly improve their lives, if they could afford the cost. Similarly, small PV systems meet the lighting and entertainment needs of remote cabins, businesses and even boats in developed countries.

However, those households that desire more energy quickly become unhappy with the limited energy resources provided by small PV systems so, if they can afford to, they upgrade. This chapter discusses some elements of planning, designing and installing larger systems. Though the information presented is far from comprehensive, the purpose is to introduce readers to the overall design process and to illustrate what a few others have done.

This chapter covers the following aspects of larger systems:

- Appliance considerations.
- Larger residential systems.

- Hybrid systems (and the use of inverters, monitors and charge regulators in hybrid PV systems).
- Institutional PV systems.
- The special case of stand-alone PV water pumping systems.

Four case studies are provided to illustrate and elaborate these sections. Note that the case studies (and the downloadable spreadsheet designs) are 'models'. System designers use these models to inform decisions on the final design of the system. In the design process, no spreadsheet or design programme provides the 'answer' – this is the work of the designer and consumer working together!

Large System Load and Appliance Considerations

With larger PV system designs, it is critical that the appliance load is estimated correctly. Before designing a PV system, review all the appliances that will be used (both initially and in the future) and consider their usage patterns (see Chapter 8). This will help decide whether PV is viable, or whether you need to seek a hybrid or alternative solution.

Large system planning starts with selecting the right type of appliances. Thus, you must ask the question, 'What can and what cannot be powered by a PV system?' The answer is not straight-forward: it depends on how large a system can be afforded, what alternative sources of power are readily available and the level of complexity desired. Many appliances can be used with PV systems, but for those appliances that cannot economically be run by PV (such as cookers or hair dryers) it is necessary to select alternatively fuelled devices.

Box 11.1 Is a solar PV system the best option?

When deciding whether a larger PV system is viable, planners must decide between several options:

- To upgrade from a small to a larger stand-alone solar system.
- To add another power source (diesel generator, wind generator).
- To connect to the grid to get cheaper power.
- To add PV to an existing diesel generator system.

Selecting the best option depends on how much power is needed, how 'dedicated' the user is to renewable energy and efficiency, how much energy independence is desired, the distance to the nearest grid power or fuel source, technical capacity available to install, design and maintain the system and the funds available to make the initial investment. These are all questions that only the client household, business or institution can answer, and it is the job of the PV designer to offer the best advice possible to meet the client needs. As opposed to small systems, large PV systems must be more carefully designed and more attention needs to be paid to safety aspects.

When selecting appliances with PV systems follow these guidelines:

- Always switch to the most efficient appliances you can afford. In the long run, it is cheaper to purchase efficient appliances than to buy PV modules and batteries for energy-wasting appliances.
- For thermal (heat) loads, switch to other energy sources. PV power is much too expensive to be used to cook, dry or run heaters. It is better to use gas, wood, kerosene or even solar thermal for heating needs.
- Some appliances, such as air-conditioners, washing-machines, compressors, large pumps or X-ray machines, cannot feasibly be powered with solar power in stand-alone systems (although there are a few 24V DC air-conditioning units on the market). It is best to power such loads directly from generators (only switch the genset on when these appliances are in use). Alternatively, investigate the costs of connecting to the grid if very large appliances are required.
- Sizing and design simulation software can be very useful for larger systems (see Chapter 12).

Choosing efficient appliances

It is far cheaper to purchase efficient appliances than to purchase extra modules to power inefficient appliances. Choosing efficient appliances can cut the cost of your PV system in half. Shop around for efficient units when making appliance purchases. (There are a number of catalogues and websites that offer efficient appliances: see Chapter 12 for more information.)

- Whenever possible use fluorescent or LED lighting (see Chapter 6).
- Use laptop computers instead of desktop PCs. They use about a quarter of the energy.
- Select efficient major household appliances such as refrigerators, washing-machines, fans and pumps which have been specially designed for PV and low-energy household applications.

Appliances that should not use Solar PV

As mentioned above, eliminate any appliances that use heat elements from your system. It is rarely viable to run heating appliances from solar electricity. Electric cookers, irons, kettles and water-heaters should never be used with PV systems.

- Cook with LPG gas, wood, charcoal, biogas, kerosene or with a solar cooker. In some large PV systems, microwave cookers (a few smaller DC versions are available) can be used but, because of the cost, this is best avoided.
- Use alternative water-heating systems such as wood, kerosene, LPG or solar thermal (see Chapter 1).

Table 11.2 describes alternatives for powering 'standard' appliances that cannot be cost-effectively powered by PV.

Box 11.2 Solar refrigerators v. 'normal' refrigerators

Table 11.1 compares 'typical' AC refrigerators with super-efficient 'solar' refrigerators. Even though they are more expensive, efficient solar-designed refrigerators quickly pay for themselves because of the lower number of modules and batteries required to run them. They will also pay for themselves, albeit more slowly, in on-grid applications. Solar fridges use extremely efficient compressors and thick insulation to improve their performance.

Table 11.1 *'Solar' refrigerators*

Refrigerator type	Voltage	Size	Price (US$)	Typical Daily Energy Consumption (Ah)	Size of array required to power daily load
Typical refrigerator	110/230V AC	12 cu ft (340l)	500	300	950Wp
Solar Refrigerator 1	12V DC	9.6 cu ft (272l)	1700	25	100Wp
Solar Refrigerator 2	12V DC	12 cu ft (340l)	2000	28	100Wp

For health centres, the World Health Organization has found that solar-powered vaccine refrigerators are often more economical than kerosene-powered alternatives and has installed PV vaccine refrigerators in thousands of health clinics around the world. Standards have been developed for solar vaccine refrigerators. For more information see Chapter 12.

If you prefer not to use electric refrigerators, use kerosene or LPG powered-refrigerators. In some cases, they may be more economically attractive than PV refrigerators. Never use absorption-type refrigerators (which use a heat element) in solar PV systems.

Figure 11.1 *Top-opening solar refrigerator*

Table 11.2 *Appliances not recommended with normal PV systems*

Appliance not recommended for PV system	Suggested alternative
Standard Refrigerator/Freezer	• Choose super-efficient fridge (see Box 11.2) • Choose LPG/kerosene fridge
Iron/Ironing box	• Use charcoal iron • Use smallest electric iron available • Run electric iron from generator when generator is 'on'
Electric cooker	• Switch to LPG cooker • Use biomass option (wood, charcoal, agricultural waste, biogas) • Use kerosene cooker
Air-conditioning	• Low energy fans (DC available)
Microwave	• Use LPG cooker or use biomass option • Run microwave from generator when generator is 'on'
Water-heater	• Use solar water-heater • Use biomass/firewood-powered water-heaters • Use kerosene/gas water-heater
Washing-machine	• Use super-efficient electric washing-machine (with hot water feed) • Use manual washers • Run washing-machine from generator when generator is 'on'
Dryer	• Use efficient electrically driven gas-heated dryer • Run dryer from generator when generator is 'on' • Use 'solar' drying or line drying
Electric heaters	• Use kerosene, LPG or wood-fired heater

Small Institutional PV Systems

There is ample experience of solar PV being used in small off-grid institutions. For example, 200Wp can provide basic light in four classrooms for night-time study at a secondary or primary school, or for the examination rooms in a rural clinic. 1000Wp can light four to six buildings, as well as a few staff houses. Hundreds of schools and clinics use solar power for night-time lighting and small appliance power.

Institutional off-grid PV systems differ from small solar home systems in several ways. Firstly, they usually demand more energy (sometimes their energy demand is too large for a simple stand-alone PV system and requires hybrid solutions, as in the clinic described in Case Study 1 below). Secondly, institutional off-grid PV systems require more careful management practice for lighting and appliances than solar home systems. Failure in institutional systems is often caused by poor operation and management practice, not poor system design.

For the above reasons, institutional systems:

- should always be properly sized and designed;
- should always select lighting and appliances carefully; and
- should always be properly managed, serviced and maintained.

Case study 1: off-grid secondary school

This example shows how a solar electric system in a small school can enable lighting systems, laptop computers and audio-visual systems to serve the school.

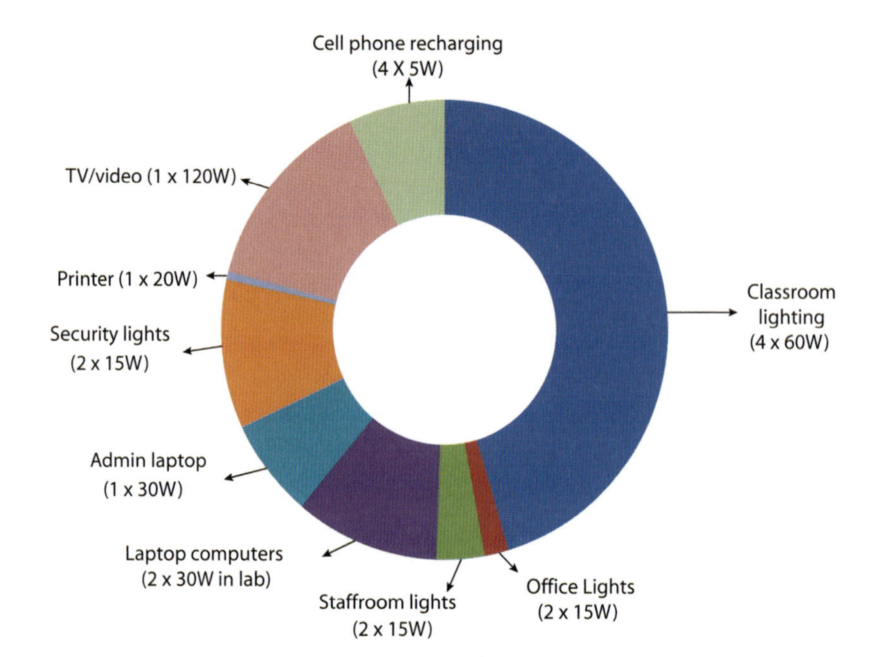

Figure 11.2 *Pie-chart of energy use in Karamugi Secondary School system. Total energy use: 1600Wh/day. AC losses: 130Wh/day. DC losses: 270Wh/day.*

This school's PV system was installed in the 1990s. The school was 5km (3 miles) away from the nearest grid connection point and the cost of extension of the grid would have been about US$50,000, which the school could not afford. The school's board compared the cost of the solar-lighting system with a generator and found that the PV system would be cheaper, calculated over a three-year period.

Table 11.3 *Karamugi Secondary School system loads*

DC Appliances Powered by PV System	
Classrooms 1–4	Four classrooms are used for evening study periods each night. Each classroom has 4 × 15W tube lamps with reflectors. Note that each classroom is painted white to increase illumination.
Office and staffroom lights	Several compact fluorescent lamps illuminate the school office and staff room.
Laptop computers	The school lab has two 30W laptop computers which are shared by students. The laptops utilize a DC-DC converter (and do not need an inverter).
Admin laptop	One laptop is used by the school administrator.
Security lights	Two fluorescent security lights are lit each night in the walkway outside the classrooms.
AC Appliances Powered by the PV System	
Printer	In practise, the bubble-jet printer is used less than half-an-hour per day.
TV/video	The TV/video is used for educational programmes 1–2 hours per day.
Cell-phone recharging (4 × 5W)	Eight staff phones use the school's charging system.
Appliances Powered by Alternative Sources	
Cooking and heating	School cooks with institutional wood stoves.
Water supply	School has a gravity water supply.

The total appliance load is about 1600Wh per day. Including losses, the total load is slightly over 2kWh per day. Note from Figure 11.2 that lighting (for evening study sessions) is the school's major energy load.

System components

Table 11.4 *Karamugi Secondary School system components*

Component	Description
System voltage	The system is configured at 12V DC and most loads are 12V DC. There is a small inverter which powers several 240V AC loads.
Array	The array is sized at 630Wp (7 × 12V DC modules). Modules are mounted on two separate pole mounts.
Batteries	Four 6V traction-type flooded-cell batteries are used in the system wired at 12V (see Figure 11.3 for configuration). Each battery has a rated capacity of 350Ah. The batteries are kept in a closet next to the office (in the same room as the charge controller).
Charge controller	The system utilizes a 60A MPPT (Maximum Power Point Tracker) charge controller. The charge controller incorporates float, boost and equalization features into its regular charging cycle. All 12V loads are connected to the controller which has a low-voltage disconnect.
Inverter	The system uses a 200W sine-wave inverter to power the AC TV, printers and phone-chargers.
AC and DC circuits	There are two separate circuits in the system. Two 240V AC power sockets run from the inverter. All lighting circuits are DC and powered directly from the charge regulator.

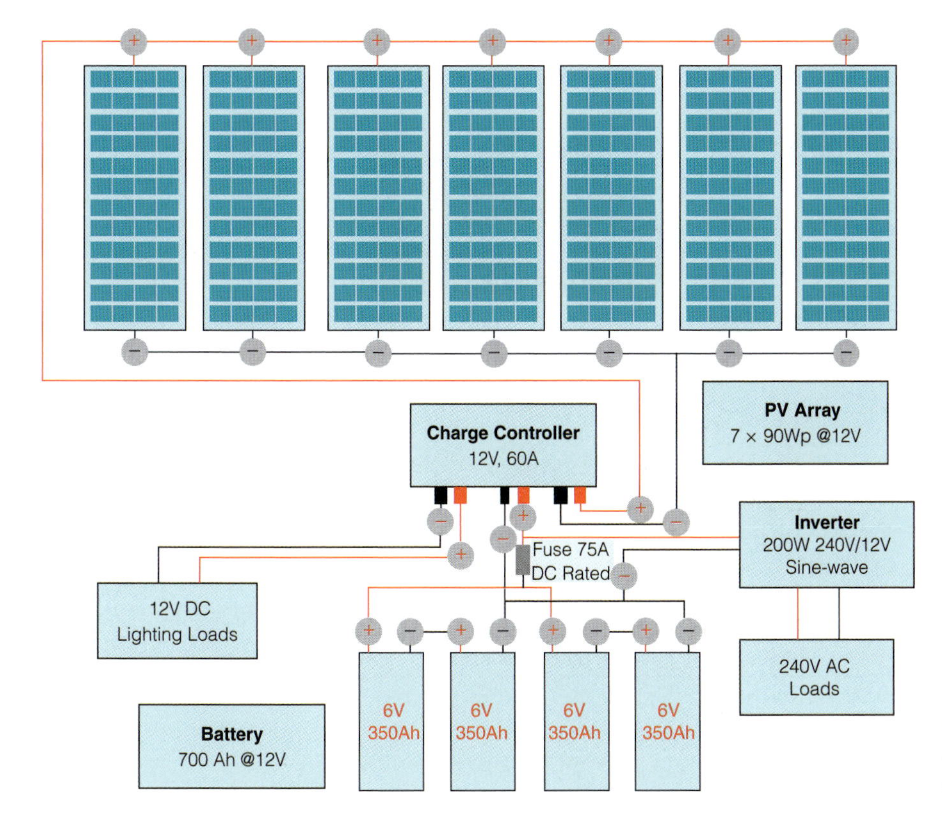

Figure 11.3 *Karamugi Secondary School system schematic*

Larger Residential Systems

Planning larger off-grid residential systems requires a focus on the needs of the people living in the house! Residential systems often work better than institutional or community systems because users have selected solar PV themselves and are fully committed to them. Planners should always work closely with system owners to ensure they understand the PV system they will be managing.

Firstly, planners should ask if the system is a full-time or a part-time leisure residence. Obviously, a full-time live-in residence will require a different design approach than one lived in during weekends and holidays only. This is especially true with regard to energy storage and cloudy season needs.

Secondly, off-grid households must review their appliance selections carefully. Review the section above and Table 11.2 when planning for the following:

- Cooking, heating and kitchen appliances.
- Refrigerators and freezers.
- Washing-machines and driers.
- Water pumping.
- Water heating.

Thirdly, think about how circuits will be designed. If there are 230/110V AC appliances, it is important to design an AC circuit and to consider whether or

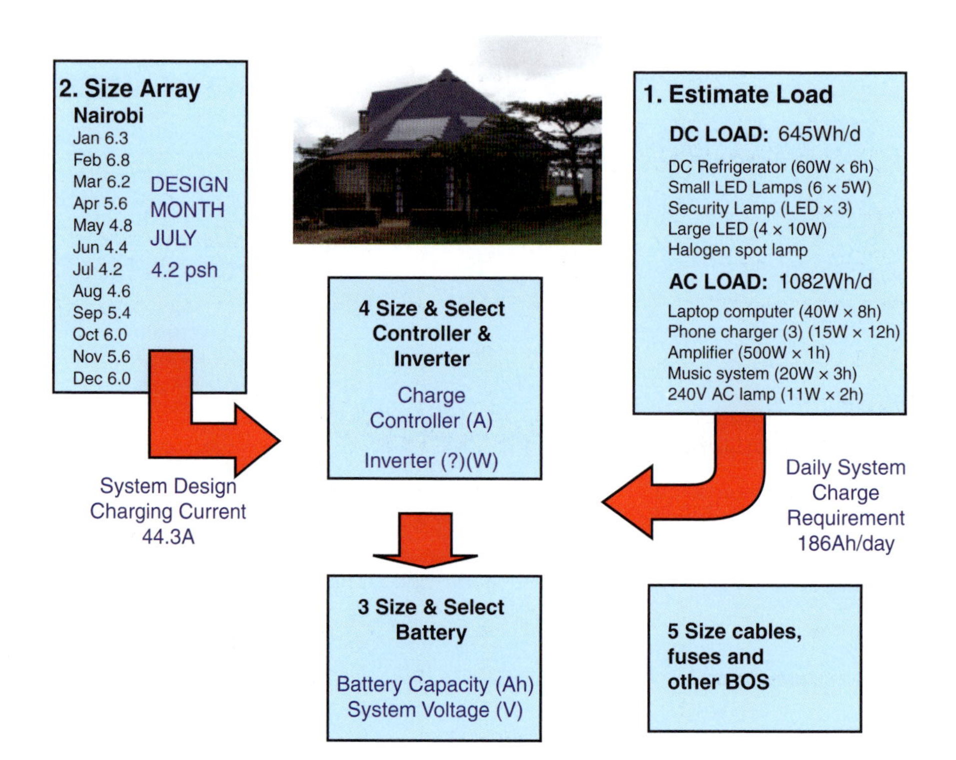

Figure 11.4 *Olonana design process*

not a DC circuit is part of the system. Inverters or supplementary power sources (as discussed below in the section on hybrid systems) may be important in large residential systems.

Case study 2: off-grid residence (Olonana 500Wp house)

The author designed the Olonana cottage as a unique off-grid weekend residence that can be used regularly for writing work, family outings and as a studio for practice sessions with musicians. Figures 11.4 and 11.5 provide information on the design process and show some of the system components.

Solar was selected for this system because there is no intention to connect the cottage to the nearest grid-line, located about 1.5km (1 mile) away. The system showcases how solar can work with an elegant design.

Olonana system loads

As can be seen from Figure 11.5, the Olonana system loads are equally divided between the lights, the refrigerator, the laptop and the band amplifier (the last of which is only used once or twice per month). On a heavy day, the total load demand is 1700–1800Wh (not including losses). Note, however, that the system is primarily used on weekends and holidays, so the average daily requirement is much lower. There are both AC and DC loads (most lights are DC).

Table 11.5 below summarizes energy systems at the cottage. Note that cooking, water-heating, water-pumping and drying do not use the PV system at all.

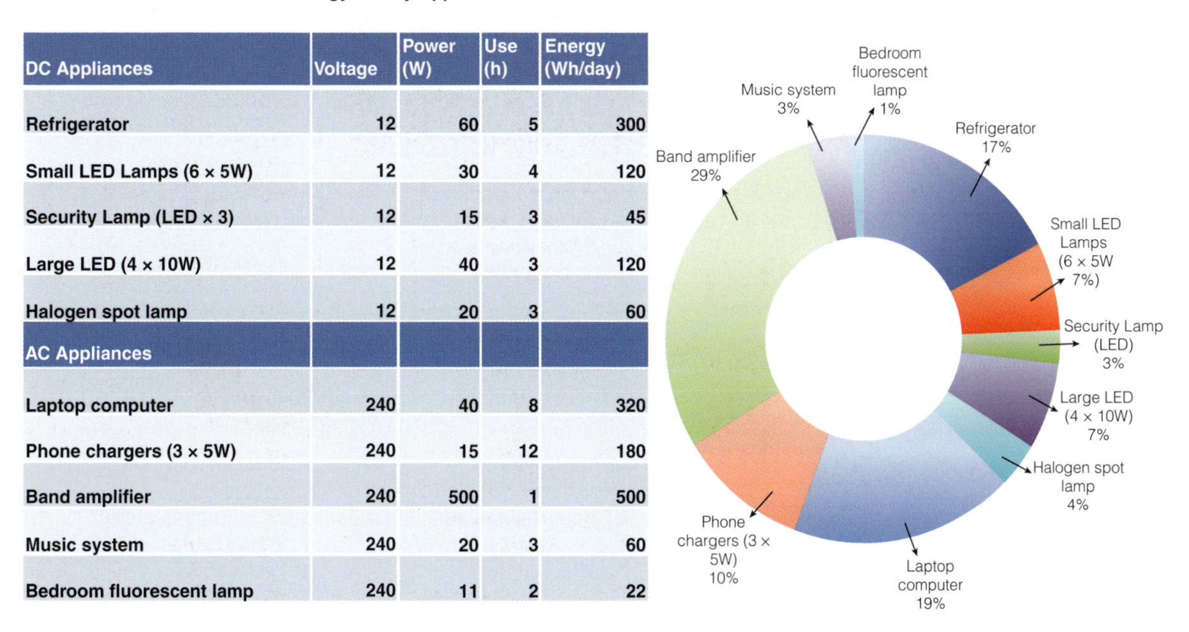

Olonana Energy Use by Appliance

DC Appliances	Voltage	Power (W)	Use (h)	Energy (Wh/day)
Refrigerator	12	60	5	300
Small LED Lamps (6 × 5W)	12	30	4	120
Security Lamp (LED × 3)	12	15	3	45
Large LED (4 × 10W)	12	40	3	120
Halogen spot lamp	12	20	3	60
AC Appliances				
Laptop computer	240	40	8	320
Phone chargers (3 × 5W)	240	15	12	180
Band amplifier	240	500	1	500
Music system	240	20	3	60
Bedroom fluorescent lamp	240	11	2	22

Figure 11.5 *Pie-chart of energy use in Olonana system*

Table 11.5 *Olonana energy and electricity use*

Energy requirement	Option Selected
Appliances Powered by PV system	
Refrigeration	Super-efficient electric fridge.
Lighting	All lighting from LEDs and fluorescents. One halogen spot light.
Band amplifier	Cottage is used for band practice once a week for four hours.
Computer	Laptop only.
Music/entertainment	Small music system.
Communication	Cell-phone chargers.
Appliances Powered by Alternative Sources	
Cooking	Biogas and LPG cookers used. Solar cooker used occasionally.
Water-heating	Biogas/LPG.
Washing clothes	Done manually.
Drying	'Solar' drying and line drying.
Heating	Fireplace meets minimal heat requirements.
Water-pumping	Rain-water catchment in 30m^3 (1060ft^3) tank. Small gravity tank used for water pressure. Water is pumped manually using foot-pump. DC PV pump to be installed.

Table 11.6 *Olonana system components*

Component	Description
Array	500Wp of flexible solar shingles incorporated into the roof of the house produce well over 2kWh of electricity on sunny days. There are five strings of six modules (each module is 17W, see Figure 11.6), and each string is wired in series at 78V open circuit and connected to an MPPT controller. The author selected solar shingles because of security concerns (solar modules would have been targets of theft) and because a roof-integrated array is attractive.
Batteries	Four 6V traction-type flooded-cell batteries are used in the system and they are wired at 12V (this means two sets of 2 × 6V batteries are wired in parallel). Each battery has a capacity of 350Ah (C20 rating). The batteries are kept in a stone shed on the outer wall of the cottage.
Charge controller	The system utilizes 2 × 30A MPPT charge controllers. One of them is wired to array strings 1–3 and the other is wired to array strings 4–5. MPPT charge controllers were selected because: they are much more efficient in harvesting power from the array; and they convert a high array voltage (78V) to the battery voltage (12V DC). The charge controller incorporates float, boost and equalization features into its regular charging cycle.
Inverter	The system uses a 1.5kW sine-wave inverter to power the AC loads in the system. These include music equipment, laptops and cell-phone chargers.
AC and DC circuits	There are two separate circuits in the system. All power sockets run from the inverter are 240V AC. All lighting circuits are DC. The refrigerator is also DC and has its own DC plug.
Battery monitor	The system has a battery monitor that provides full information about the battery state of charge, energy use (i.e. amp hours removed or added) and need for equalization.

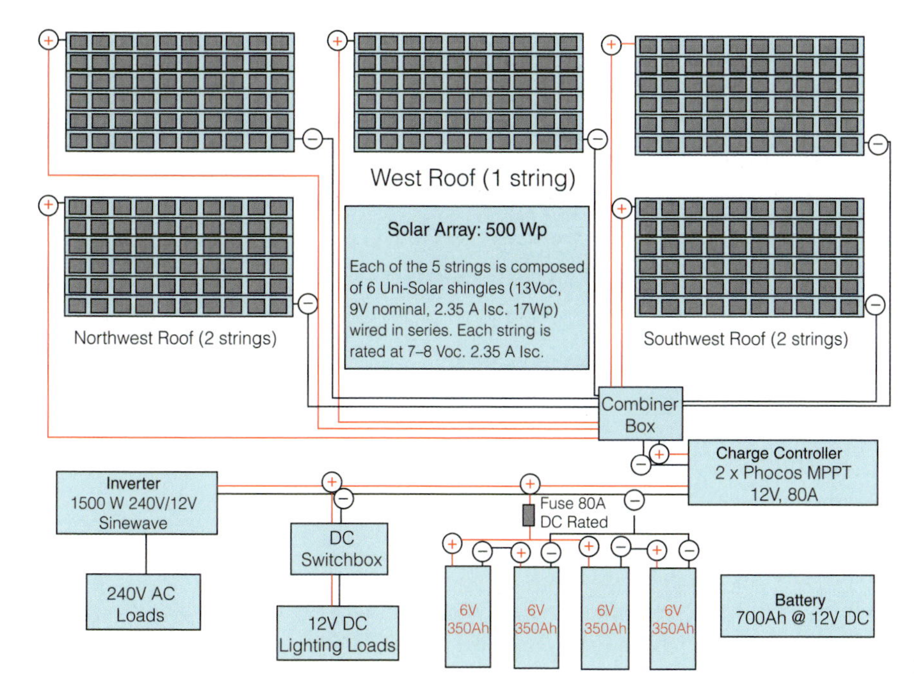

Figure 11.6 *Olonana system schematic*

Hybrid systems

Hybrid systems combine solar electric, generators and, occasionally, other renewable energy systems to increase energy availability. In hybrid PV systems, the solar array charges the batteries and supplies primary daily power requirements. A generator is used to charge batteries when there is not enough sunshine (or wind), to run battery equalization and to power specific large loads that the PV system cannot power. Hybrid systems are best used in remote locations where solar or wind resource is variable and where there is occasionally a need for large amounts of power. Most are PV-diesel hybrid systems.

In general, stand-alone PV systems are economically viable without gensets when energy requirements are less than about 5kWh per day and there is good constant sunshine. As energy needs increase, hybrid systems with diesel gensets and/or wind generators begin to look more attractive. Hybrid systems allow more flexibility in energy end-use because they ensure that power is available from a variety of sources. For example, during cloudy periods, a system operator can use wind or diesel generators to charge batteries and not be 100 per cent reliant on PV.

Cost is not the only factor to be considered when designing a PV system and there are other questions customers should ask:

- How far must fuel be carried and how often must trips be made to get it?
- Do people want to deal with the noise and fumes of petroleum generators?
- Is self-sufficiency or 'eco-friendliness' a consideration?
- How reliable are the alternatives?

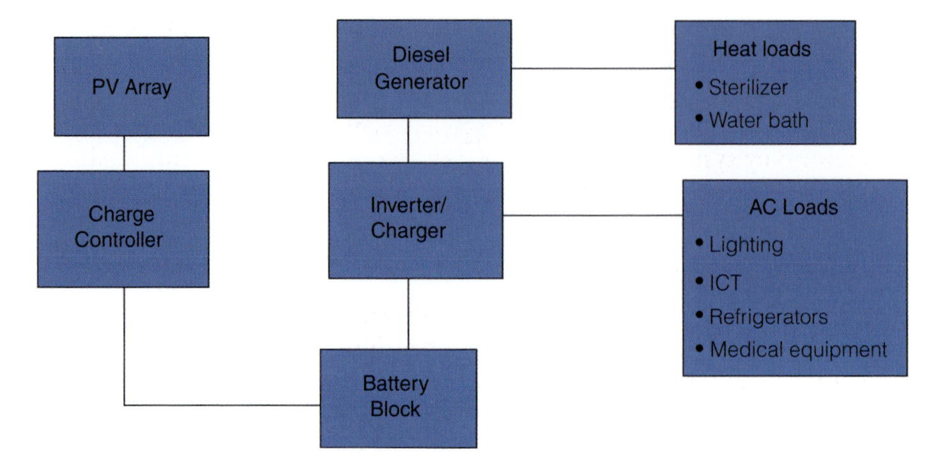

Figure 11.7 *Schematic diagram for a hybrid system*

Hybrid system parts and operation

Figure 11.7 shows a schematic diagram of a hybrid system (without fuses, switches, etc.). Key parts of hybrid systems include the following:

- Generator. As explained below, the generator is used to provide additional power to the system.
- Inverter-chargers. In addition to converting low-voltage DC battery power to AC power for appliances, inverter-chargers convert AC power from the generator to DC power used for charging the batteries. It is important to select the right type of inverter-charger.
- Switching and monitoring equipment. With hybrid systems it is important to have additional switching equipment. This should be carefully designed for the needs of your system.

Back-up generators

Generators are used for three primary purposes:

1. As a back-up source for charging and topping-up when there is not enough sun.
2. As a way to equalize batteries.
3. As a power source for large loads which cannot be powered by the PV system and inverter.

Generators as back-up power sources

For large systems, powering 100 per cent of the load with solar is often expensive because the array and battery must be oversized to meet demand during the cloudiest months or during periods when demand for energy is heavy. To get around this, many designers plan so that solar meets 75 to 90 per cent of the total energy requirement. During cloudy periods (or when days are short), a generator is used to meet energy shortfalls and to top up the battery when it falls to a low state of charge. This can reduce costs of PV arrays and batteries considerably and extend battery life. Many PV-diesel hybrid systems arise when batteries, an inverter-charger and then PV is added to an existing generator-only system – the amount of time the generator is run for is reduced, which also reduces fuel consumption, noise and fumes.

Generators for equalizing batteries

As mentioned in Chapter 4, equalizing is important for large flooded-cell battery banks, especially during periods of low sunshine when PV arrays may not produce enough power to equalize battery banks (even if the charge controller has a setting for this purpose). Generators are useful tools to produce the minimum current required for this task (usually C10 to C20, see Chapter 4).

Generators as power sources for large loads

Sometimes provisions for occasional large loads are included in systems that cannot be powered by the inverter or PV system. For example, in off-grid hospitals, this might include X-ray machines or electric sterilizers. Other large loads not ordinarily powered by PV or inverters include workshop equipment such as welders, air-conditioners, borehole pumps or compressors. Connecting such loads directly to the generator, and running them from the generator when needed, can greatly increase the overall flexibility of the system. As an example, the clinic in Case Study 3 below runs the generator three times a week to power the sterilizer and laboratory HIV-testing equipment.

Selecting inverter-chargers

Inverter-chargers are complex devices and before selecting one the designer should consult the manual to make sure its specifications are those required. It has two main functions: to provide AC power and to charge the battery. So, the AC output needs to be the correct wattage and AC voltage to power the loads, and its battery-charging function needs to provide sufficient current to charge the batteries (which can be adjustable) and be of the correct DC voltage. Inverter-chargers can also be powered by grid-power in back-up systems where the grid is unreliable.

Selecting back-up generators

In many hybrid systems the diesel generator will already be on-site, but this is not always the case. An existing generator needs to be checked to ensure that it is suitable for the system, and if a system is being designed from scratch then the right back-up generator must be selected carefully and with expert advice. The generator must be sized as part of the overall system design.

Table 11.8 lists some of the key features to be considered in hybrid system generators.

Table 11.7 *Key factors in inverter-charger selection*

Inverter-charger specification	Notes
AC output power (W) and AC output voltage	Enough power to power all AC appliances on circuit? Correct AC voltage? 120 or 240V AC?
AC input power (W) and voltage	Usually the same as the AC input power (W) and voltage but needs to be checked. Is it suitable for the type of generator in the system? Correct voltage?
Battery-charging current and other battery-charging specifications	What is the maximum DC charging current that can be delivered to the batteries? How long will this take to charge the batteries (assuming the batteries are in a low state of discharge)? Can the battery bank accept the maximum charge? What is the correct DC voltage for the battery bank – 12, 24 or 48V DC? Will it automatically provide the battery bank with an equalizing charge? Is it suitable for charging sealed batteries? The potential output (battery charging ability) of other power sources (e.g. wind) should be considered when selecting.
Integrated charge controller	Does the unit have an integrated solar charge controller? This is not necessary but it is an option worth considering.
Remote monitor	Is a remote monitor for the inverter-charger available? These are really important in large systems as they enable the system manager to see what is happening without having to go into the battery room.
Automatic start for generator	Will the inverter-charger automatically start the generator when the batteries get too low – or does this need to be done manually? The inverter-charger manual will provide details.

Remember, depending on the situation, a 100 per cent PV-powered system might be desired or a 100 per cent generator-powered system might be desired (or anything in between). Table 11.9 presents a number of indicators that may help you choose between hybrids, generators or solar electric alternatives.

Figure 11.8 *Typical standby generator*

Table 11.8 *Generator features*

Generator feature	Notes
Engine speed	The slower the engine speed, the longer the generator is likely to last, the more quietly it will run and the less fuel it will consume (1800rpm is a considered a good speed).
Cooling system	Liquid-cooled engines are preferable because they run quieter and last longer.
Maximum load of the selected generator	Generators normally run best at a load that is 80–90% of their rated load. The rated capacity of the generator (in kVA) is not the actual normal running output.
Fuel type	Generators can run on petrol, diesel, propane (or LPG), biodiesel or biogas. Diesel is usually preferred over petrol (gasoline) for safety and efficiency reasons, though it produces more smoke pollution.
Fuel storage	Safety, temperature and security are all considerations.
Earthing/grounding	Generator earthing methods varies by region and type of generator. Follow recommendations of manufacturers and local codes.
Generator starting	Consider whether the generator has an autostart (which means that the inverter can automatically turn on the generator when required), a remote start (meaning it can be turned on from a central control board) or a manual start.
Normal temperature and altitude of site	Higher temperature and elevation above sea-level can reduce the overall performance of generators. A generator at 1500m (roughly 5000 feet), for example, would be expected to operate at 82 per cent of its rated output at sea-level. See generator instructions for de-ratings.

Table 11.9 *Comparing generator and solar electric power*

Situations which favour Generators	Situations which favour Solar Electricity
Power requirements are above 5kWh/day	Power requirements are below 5kWh/day
Fuel supply is stable	There are frequent shortages of fuel
The fuel depot is nearby	Fuel depots are far away
Fuel and maintenance costs are low	Fuel and maintenance costs are high
Cloudy weather is common	There is high solar radiation (5 or more peak sun hours throughout the year)
Low-voltage lamps and appliances are not available	The equipment is expected to last a long time with little maintenance
Deep-discharge batteries not available	Noise and exhaust fumes must be avoided
Light and appliances are required for a short time only each day	Lights and appliances are in constant use
Power demand varies greatly	Power demand is steady

Case study 3: a hybrid clinic system

Off-grid solar PV systems often power clinics in remote parts of developing countries. The Kamabuye clinic, in Rwanda, serves several thousand inhabitants of an off-grid community. It is an example of a site where a wide variety of loads required a hybrid PV design.

A hybrid system is used because there are several loads that cannot be powered by a PV/inverter system (most notably the sterilizer) and because there are several months where cloudy weather limits PV array output.

Table 11.10 below provides information on appliances used in the centre.

Table 11.10 *Kamabuye energy and electricity use*

Energy Requirement	Option Selected
AC Appliances	
Lighting: Admin. block, maternity, ward, security	LED and compact fluorescent.
Cell-phone recharging	Staff phones recharged every day.
Centrifuge	Used twice a week.
Laptop computer	Used daily.
Printer	Used occasionally.
Copier	Used daily.
Refrigerator	Used daily. Super-efficient type required.
Video/TV	Used twice a week for out patients and education.
Generator-run Appliances	
Sterilizer	3kW unit used to sterilize all instruments.
HIV testing	Used twice a week when screening for HIV.
Non-electric Applications	
Cooking and hot water	Water is heated using gas and wood stoves.
Drying	Clothes are dried outside on lines.

Table 11.11 *Kamabuye system components*

Component	Description
System voltage	Kamabuye has a system voltage of 12V DC, meaning its batteries and solar array are configured at 12V DC.
Array	During the design month (October), the lowest levels of insolation are 4.71kWh/m^2/day. However, because normal electricity use is lower when the health centre is not at full capacity, and funding is constrained, the solar PV system was based on a slightly higher figure of 5kWh/m^2/day. The array installed contains 12 $\times$ 90Wp modules, each of which provides 5.3A in normal conditions. The 12 modules are mounted on a fenced rack behind the health centre.
Batteries	Flooded-cell stationary batteries were selected. Design criteria called for 1312Ah of batteries at 12V DC (2 reserve days, with a maximum allowable DoD of 50%). A slightly smaller set of 6 large 2V cells rated at 1200Ah was installed. The batteries are housed in a storage room used exclusively for the batteries, charge controller, inverter and control board.
Charge controller	In this system, the main function of the charge controller is to regulate charge from the array and prevent over-charging. The selected MPPT charge regulator is rated to accept 80A from the array. It also incorporates a monitoring function that tells the state of charge of the battery. Note that the charge regulator in this system does not have a low-voltage disconnect, as all loads come from the inverter.
Inverter-charger	The inverter-charger was selected based on the maximum possible demand of the 240V AC appliances and the charging needs of the battery bank. The unit selected can provide 2000W power when all appliances are turned on.
Generator	The clinic has a 10kVA diesel generator. This generator powers the sterilizer and HIV-testing equipment directly three times a week. It is also used to top-up the battery during cloudy periods (e.g. October) and to equalize the battery once every three months.
AC and DC circuits	The array and battery are wired at 12V DC. All load circuits are at 240V AC. There are two main circuits: the 'low power' loads that originate from the inverter and the 'high power load' (sterilizer) that is powered directly from the generator.

Table 11.12 *Powering appliances in remote clinics*

Appliances PV can easily power	Appliances PV cannot power inexpensively
Microscopes	X-rays (use a generator)
Centrifuges	Sterilizers (use gas, kerosene or solar concentrator)
Theatre lights	Laboratory equipment that uses heat elements (use water bath)
Efficient refrigerators	
Computer systems	
Communications (HF radio)	
Suction pumps	
Surgical drills	

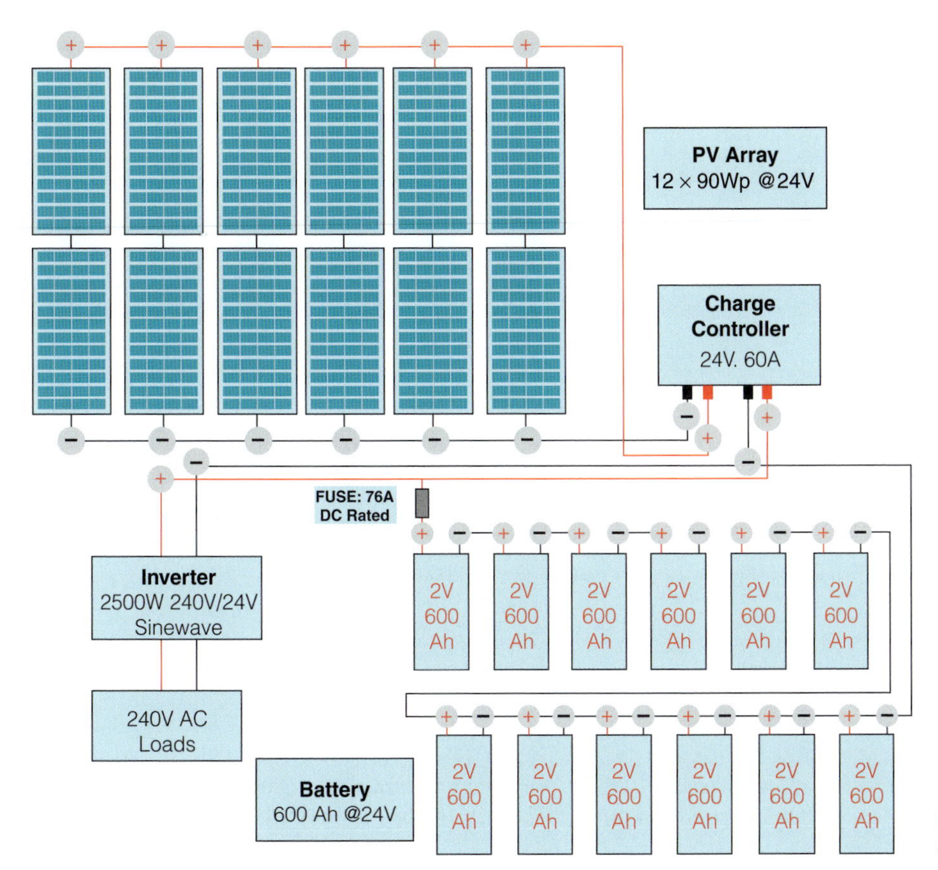

Figure 11.9 *Kamabuye system schematic*

PV Water-Pumping

PV pumping systems are often attractive options for small-scale community water supply, household water supply, agriculture and livestock needs (PV is usually not an attractive option for medium and large-scale irrigation). This section gives an overview of solar water-pumping system planning processes.

As shown in Figure 11.10, the process of designing solar-pumping systems is significantly different from that outlined in Chapter 8.

- Remember, solar water-pumping is a specialized field. Always consult an expert on solar water-pumping when selecting your pump and designing your system.
- Get the right type of pump and motor. Depending on your situation, there are various types of pumps (diaphragm, piston, centrifugal, helical rotor, etc.) and motors to meet your needs. Find out the right type for your situation.
- Buy a whole system, not components. Solar-pumping system components are designed to work together, and putting together pieces from different manufacturers may lead to disaster!
- Various pumping alternatives should be considered carefully before choosing solar. Wind- and diesel-powered pumping systems may be more economical or practical in many situations. Solar power is attractive in situations where demand is too high (or the well is too deep) to be supplied by hand-pumps.
- Solar-pumping sites should have good security – theft of modules is a common problem.

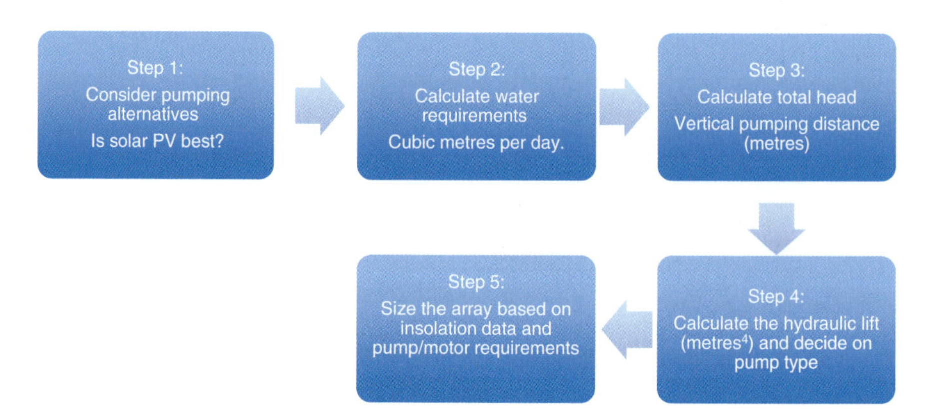

Figure 11.10 *Designing solar PV pumping systems*

Deciding to use a PV pumping system

Solar water-pumps require high, consistent sunshine. The most attractive sites for PV pumping are where less than 30m^3 (1060ft^3) of water are required per day, the head is less than 100m (330 feet) and there is no grid electricity. Solar is especially cost-effective where petroleum costs are high (e.g. due to the remoteness of the site) and where wind resources are limited. Most solar-pumping systems avoid batteries by pumping water and storing it in an overhead tank.

To make an intelligent choice on whether to use a PV pumping system, the following data is required:

- The water requirements of the site (m^3 of water required).
- The depth of the water source (m).
- The water source (water quality, borehole yield).

- The solar radiation availability.
- Information about other available technology alternatives.

Step 1: consider all pumping alternatives and estimate their costs

When considering solar-pumping, bear in mind the following:

- Diesel pumping sets can deliver water in far larger volumes and from greater depths than PV, and are usually a better choice for medium and large-scale agriculture.
- Hand-pumps or foot-pumps are often more cost-effective than PV for shallow wells and for small volumes of water.
- Wind-pumps deliver similar volumes of water as PV pumping systems, but are more secure (they cannot be stolen or vandalized) and can deliver from greater heads. In windy locations, always consider wind-pumping as an option!

A good way to compare solar-pumping with alternatives is to calculate the lifetime cost of each system in terms of cost per delivered cubic meter. See Figure 11.11.

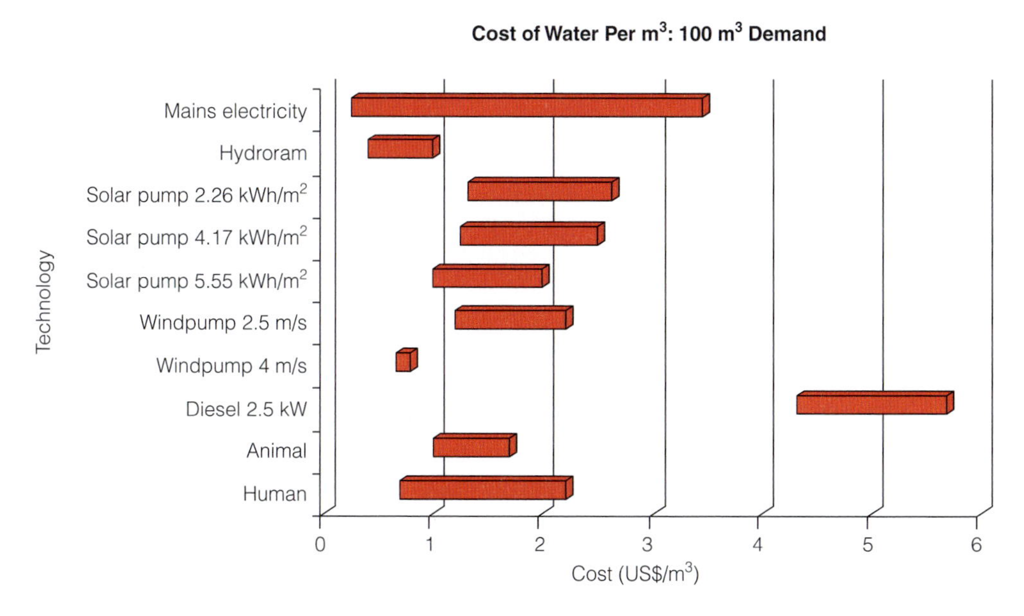

Figure 11.11 *Comparing water-pumping costs (but bear in mind that costs vary from location to location)*

Step 2: calculate water requirements

First, calculate the volume of water required per day by the household, community, livestock or farm. Table 11.13 provides a basis for estimating the likely water requirement for a number of needs. To make a calculation, estimate the number of people and animals in a settlement and calculate the total in cubic metres per day.

Table 11.13 *Water requirement guide for humans and livestock*

Water use	Typical daily water requirement (litres/day)
Human	20–40 (rural standpipe in developing country) 200–300 (piped water)
Cattle (beef)	40
Cattle (diary)	80–150
Sheep/Goat	5–10
Donkey	20
Pig	15
Camel	20
Chicken (per 100)	15

Step 3: calculate the vertical pumping distance (head)

The depth of the well/borehole plus the height to which it will be pumped is called the 'head'. The pump will have to push the water this full distance. A shallow well might have a head of 8–20m (25–65 feet), while a borehole might have a head of 60m (200 feet) or more. Solar electricity is expensive for use in deep borehole applications and is not normally considered where water is more than 150m (500 feet) below the surface.

Step 4: calculate the hydraulic lift and decide on type of pump

The amount of water needed (in cubic metres, m^3) multiplied by the total head gives the hydraulic lift (m^4). This number is used by engineers to select and size the pump and to estimate the amount of power needed to pump the water (i.e. the size of the array).

Figure 11.12 shows several common PV pumping configurations. In addition to the hydraulic lift, the water quality, the need for treatment, the borehole yield and the type of storage play a large role in determining the type of pumping system. Always consult an experienced expert when deciding on the pumping system! See also Chapter 12 for resources on solar-pumping.

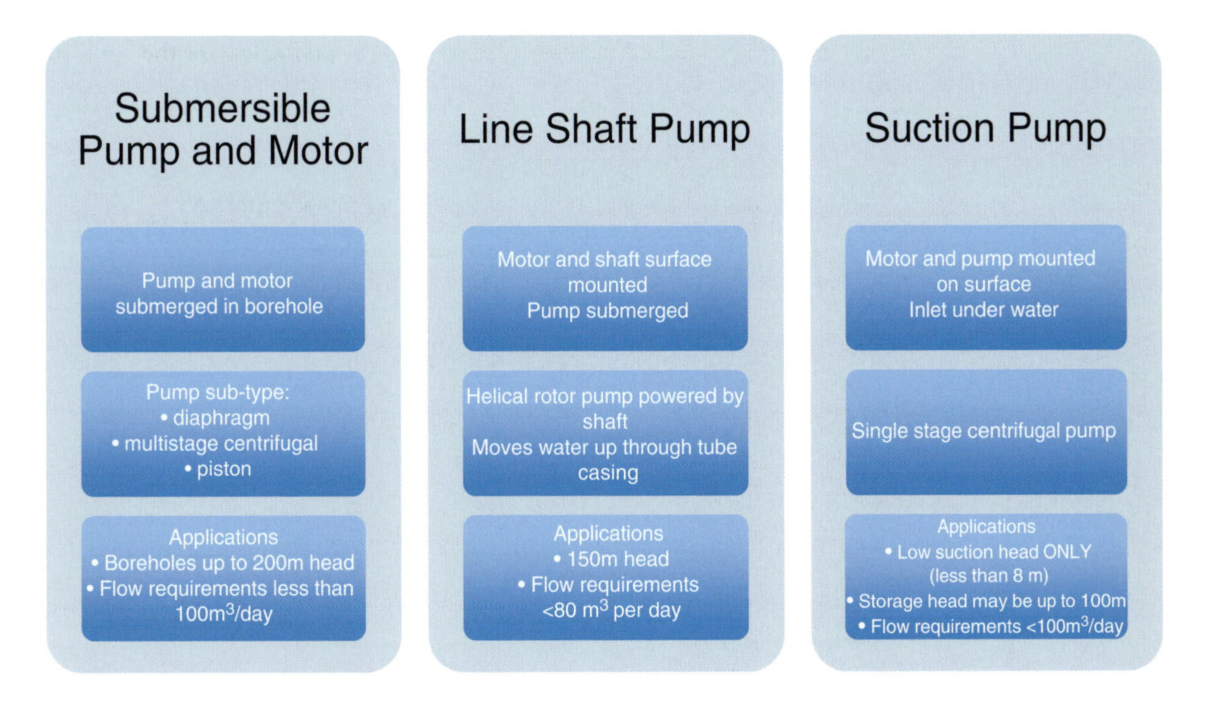

Figure 11.12 *Common pumping configurations*

Step 5: size the solar array

The size of the array selected for a pumping system largely depends on the type of motor and pump. Most solar-pump packages have standard methodologies that they use to size solar arrays for their pumps that are based on the hydraulic lift required and the design insolation value. A solar-pump supplier will be able to help you size the array for the pumping system you need (see also Chapter 12).

Consider other factors

- Borehole yield. Do not exceed the daily yield of the bore hole.
- Water condition and treatment needs. Have water quality checked by an expert.
- Site security.

Solar PV Pumping Case Study

This case study features a desert settlement with 300 people, 500 cattle and 2000 goats. The water source is a shallow well.

Step 1: consider all pumping alternatives and estimate their costs.

Wind resources at the site are poor. Because of its isolation and the costs, a diesel pump is not desired. Solar resources are high (5–6kWh/m^2/day).

Step 2: calculate water requirements. About 40m^3 of water per day is required.

People: 300 × 40 litres/person/day = 12m^3

Cattle: 500 × 40 litres/cow/day = 20m^3

Goats: 2000 × 7.5 litres/goat/day = 15m^3

Total requirements: 47 m^3/day (with 3m^3 added for growing needs this equals 50m^3)

Step 3: calculate the vertical pumping distance (the head). This is 15m.

Step 4: calculate the hydraulic lift and decide on the type of pump. The settlement would require a total hydraulic lift of 750m^4 per day (15m × 50m^3). A submersible pump could be used.

Step 5: size the solar array. The radiation at the site is 5kWh/m^2/day; approximately a 1.2kWp array would be required (calculated using the solar-pump supplier's method in catalogue).

With the above information it is possible to speak to suppliers about possible options and to investigate what is generally on the market to suit the systems needs.

12

Off-Grid PV and Solar Energy Resources

This chapter provides resources for those interested in gaining further information about solar products, training opportunities and general knowledge about solar. Please note that the section is largely based on the author's network and experience. It is not meant to be a comprehensive list. Also note that the author and publishers cannot be held responsible for the quality or content of the information contained in these sources or the products recommended.

Introduction

The information contained in this book is basic and only meant to be a starting point for solar do-it-yourselfers. Nevertheless, many people bitten by the solar bug may want more information – or help in solving problems as they arise.

Even though solar electric technology is growing at an incredible rate, the field of off-grid solar electricity is still comparatively small when measured against grid-connected. Therefore, getting products for, and information about, small solar electric systems for off-grid purposes is still not as easy as it might be. In fact, many people are confused by the differences between 'off-grid' and 'on-grid' PV.

Fortunately, there are many champions in the field of off-grid solar power. This chapter collects information about some of them. Important sections covered below include:

- Sources of information including magazines, websites, books about solar electricity and off-grid living.
- Places where you can find further training or locate expertise.
- Sources of information about solar energy resources, measuring tools and design programmes.
- Companies that specialize in off-grid solar energy and renewable energy equipment.

Because of the author's particular experience, this chapter focuses on European and North American sources of information. Nevertheless, there are strong solar energy communities in China, India, and elsewhere across the globe. No region holds a monopoly on solar energy information or products, so readers should seek out national groups in their own countries for more information.

General Sources of Information

Magazines

Home Power Magazine (USA)

A bi-monthly magazine that has been a primary source of information and expertise among American off-grid solar energy systems installers for over two decades. For the beginner and the expert, it provides a magazine and on-line content with a wealth of information about installation, design practices and equipment. It also has CDs with back issues that are extremely useful for the hands-on practitioner. www.homepower.com

Sun and Wind Magazine (Germany)

A good general magazine about solar and wind energy with content that ranges from market information to hands-on stories. Covers global developments in renewable energy. www.sunwindenergy.com

Renewable Energy World (USA/UK)

A wide-ranging magazine about the global developments in renewable energy. Company and job information. Both print and PDF versions available. www.renewableenergyworld.com

Books

Earthscan

The following books offer information about solar and other renewable energy sources.

Earthscan

Title	Description of book
Photovoltaics for Professionals	Describes the practicalities of designing and installing photovoltaic systems, both grid-tied and off-grid – though the emphasis in the current edition is on grid-tied. Written for electricians, technicians, builders, architects and building engineers.
Applied Photovoltaics	Accessible and comprehensive guide for students of photovoltaic applications and renewable energy engineering. Step-by-step guide through semiconductors and p-n junctions, cell properties and design, PV cell interconnection and module fabrication. Covers stand-alone systems, specific purpose systems, and grid-connected systems, photovoltaic water pumping, system components and design.
Selling Solar	Shows how, at the start of the 21st century, PV technology began to diffuse rapidly in select emerging markets after years of struggling to take off. What were the initial barriers to diffusion? How were they overcome? Who did it? And how can this success be replicated? The book also describes how entrepreneurs affected profound technological change not just through the solar systems they sold, but through the example they set to both new market entrants and policymakers.

Title	Description of book
Planning and Installing Photovoltaic Systems (Second Edition)	This joint Earthscan/German Solar Energy publication covers all aspects of grid-tied PV and is excellently illustrated throughout.
Solar Domestic Water Heating (Expert series)	A comprehensive, accessible and richly illustrated introduction to all aspects of solar water heating.
Renewable Energy Engineering and Technology	Comprehensive guide on renewables technology and engineering, intended to cater for the rapidly growing numbers of engineers who are keen to lead the revolution. All the main sectors are covered – photovoltaics, solar thermal, wind, bioenergy, hydro, wave/tidal, geothermal – progressing from the fundamental physical principles, through resource assessment and site evaluation to in-depth examination of the characteristics and employment of the various technologies.
Understanding Renewable Energy Systems	Beginning with an overview of renewable energy sources including biomass, hydroelectricity, geothermal, tidal, wind and solar power, this book explores the fundamentals of different renewable energy systems. It not only describes technological aspects, but also deals consciously with problems of the energy industry. Contains a free CD-ROM resource, which includes a variety of specialist simulation software and detailed figures from the book. Ideal companion for students of renewable energy and engineering students.

The titles above can be ordered from the Earthscan website: www.earthscan.co.uk

New Society Publishers

Title	Description of book
Power from the Sun	A practical guide to solar electricity.
Photovoltatic Design and Installation Manual	Solar Energy International's training manual.
Solar Water Heating	A comprehensive guide to solar water and space heating systems.
Real Goods Solar Living Source Book	A complete guide to renewable energy technologies and sustainable living. See Real Goods catalogue.

The titles above can be ordered from the New Society website: www.newsociety.com

Other books

Title	Description of book
Photovoltaic Systems (American Technical Publications, Homewood, IL, 2007)	Developed by the National Joint Apprenticeship and Training Committee for the Electrical Industry in the US.
The Solar Electric Independent Home Book (Fowler Solar Electric, Worthington MA, 1991)	Book written for do-it-yourselfers by an American off-grid solar pioneer.

Training and development organizations

In order for the technology to be effectively disseminated and utilized, organized training efforts need to occur. These groups below have worked with installers, distributors, sales agents, community leaders and end-users.

Solar Energy International (Colorado, USA)

On-line courses and on-site design and installation courses. Some international training and development work.

www.solarenergy.org

Renewable Energy Academy (RENAC) (Berlin, Germany)

Provides courses on most renewable energy technologies, including off-grid PV, PV-wind hybrid systems and mini-grids.

www.renac.de/en/home

Lighting Africa

International finance organization dedicated to improving access to modern lighting services for low income groups in Africa. Helps in the promotion of low cost, high quality lighting products through networking, quality monitoring, policy initiatives and competitions.

www.lightingafrica.org

Solar Electric Light Fund

International development organization promoting solar power and wireless communication, with many projects.

www.self.org

Alliance for Rural Electrification

International development organization involved in rural electrification, with a strong emphasis on solar electrical systems.

www.ruralelec.org

Green Dragon Energy

Provides training courses and is a source of books on renewable energy (mainly off-grid). Has been involved in solar energy for many years.

www.greendragonenergy.co.uk

Solar Aid

International NGO promoting solar electricity access, primarily in sub-Saharan Africa. Works with volunteers and private contributions from sponsors.

www.solar-aid.org

Solar Resource Data and Simulation Software

PVGIS

A website with compiled data and links for meteorological data from sites all over the world. Includes on-line simulation tools.

re.jrc.ec.europa.eu/pvgis

NASA Surface Meteorology and Solar Energy

Database providing radiation data for all global locations.

http://eosweb.larc.nasa.gov/sse/

PV*Sol

Professional simulation programme for design and simulation of stand-alone and grid-connected photovoltaic systems. Demo version can be downloaded free.

www.valentin.de

PVsyst

Professional simulation programme for design and simulation of stand-alone and grid-connected photovoltaic systems. Demo version can be downloaded free.

www.pvsyst.com

Homer

HOMER is a computer model for evaluating design options for larger renewable and conventional hybrid and distributed power sources. Not for small systems. Great as a learning tool – and free.

www.homerenergy.com

Standards, codes of practice and policy

Codes of practices guide the installation methods, choice of materials, workmanship and system maintenance, ensuring a long and safe lifetime for PV power systems. They provide the criteria by which systems can be inspected to ensure sustainable and safe operation.

The following codes of practice and lists of pre-qualified products are available:

International Electro-technical Commission (IEC)

The IEC has developed (and continues to improve) a set of comprehensive standards for solar photovoltaic energy systems. This is done by Technical Committee 88.

www.iec.ch/zone/renergy/ren_standardization.htm

Global Approval Program for Photovoltaics (PV-GAP)

PV GAP is a not-for-profit organization that promotes the use of internationally accepted standards, quality management processes and organizational training in PV standards and quality. Products which bear the Photovoltaic Global Approval Program (PV GAP) Seal have been tested in an acceptable manner. PV-GAP has done extensive work on small off-grid PV systems.

www.pvgap.org

World Bank-assisted projects

World Bank-assisted projects have developed standards and pre-qualified equipment for small off-grid PV systems. See the following:

- World Bank Asia Sustainable and Alternative Energy Unit. www.worldbank.org/astae
- Sri Lanka Energy Renewable Energy for Rural Economic Development Project. www.energyservices.lk
- Golden Sun quality mark is issued by the China General Certification Center. For list of qualified products. www.cgc.org.cn
- Bangladesh Rural Electrification and Renewable Energy Development Project. For list of qualified products. www.idcol.org
- Bolivia Decentralized Infrastructure for Rural Transformation Project. www.idtr.gov.bo

Australian National Standards

Standards in the AS 4509 Stand-alone power systems series comprise:

- AS 4509.1-1999 Stand-alone power systems Part 1: Safety requirements.
- AS 4509.2-2002 Stand-alone power systems Part 2: System design guidelines.
- AS 4509.1-1999 Stand-alone power systems Part 3: Installation and maintenance.

This family of standards is for use by owners, designers, installers and users of stand-alone power systems. It was developed in response to concerns from the renewable energy industry and regulators over the poor design and installation of renewable energy powered stand-alone systems. For information go to www.standards.org.au.

Equipment

The list of commercially available equipment below is, as mentioned previously, not intended to be comprehensive or an endorsement of the products. It is designed to give the reader a start in the world of off-grid PV. Many of the companies mentioned supply a range of products, for example modules, lamps and charge controllers, so be sure to check all the lists when researching.

Make sure that you know what you are getting. Read the appropriate chapter before making purchase decisions. Always use products suitable for off-grid applications, not equipment designed for grid connect applications – sometimes the terminology can be confusing. If in doubt, explain clearly to the supplier the type of system the product is going to be used in.

Solar PV Modules

Solar companies are constantly changing and updating their products, and many companies have been changing their product lines to address on-grid markets only. This list contains PV companies that have a strong 12V DC product line.

Name	Products	Website
Suntech	Wide range of grid-connect and off-grid solar products	www.suntech-power.com
Sharp Solar	Wide range of grid-connect and off-grid solar products	www.sharp-solar.com
BP Solar	Wide range of grid-connect and off-grid solar products	www.bpsolar.com
Kyocera Solar	Polycrystalline modules for on and off-grid applications	www.kyocerasolar.com
Evergreen Solar	Ribbon-type solar cell modules	www.evergreensolar.com
Photowatt	Wide range of grid-connect and off-grid solar products	www.photowatt.com
SolarWorld	Wide range of off-grid and on-grid modules	www.solarworld.de
Free Energy Europe	Amorphous modules for 12V DC applications	www.freeenergyeurope.com

Batteries

Read Chapter 4 on batteries carefully before selecting the units you will use. Remember, for small systems (>100 Wp) it is often cheaper and easier to select a local modified SLI battery than to import an expensive battery from a distant location. The companies listed below supply high quality deep discharge batteries that are useful in larger solar PV systems.

Name	Country of origin	Products	Website
MK Battery Deka Solar	USA	Gel/AGM and flooded batteries	www.eastpenn-deka.com
Exide Technologies	USA, worldwide	Wide variety of lead-acid batteries	www.exide.com
EnerSys	USA	Wide variety of lead-acid batteries	www.enersysmp.com
Power Battery	USA	Flooded, AGM, gel batteries	www.powerbattery.com
Surrette	Canada/USA	Deep cycle batteries	www.surrette.com
Trojan Battery	USA	Flat plate, gel, AGM batteries	www.trojanbattery.com
Hoppecke batteries	Germany	Deep cycle batteries	www.hoppecke.com
Sonnenschein	Germany	Deep cycle, AGM batteries	www.sonnenschein.org
Elecsol Batteries	UK	Various leisure and solar batteries	www.elecsolbatteries.com

Charge controllers and inverters

The list below provides a basic introduction to major inverter and charge controller suppliers in the off-grid market.

Name	Country of origin	Products	Website
Phocos	Germany	Charge regulators, inverters, lights	www.phocos.com
Apollo Solar	USA	Charge controllers, inverters	www.apollosolar.com
Studor	Switzerland	Inverters, monitors	www.studer-inno.com
Blue Sky Energy	USA	Charge controllers	www.blueskyenergyinc.com
Morningstar	USA	Charge controllers, inverters	www.morningstarcorp.com

(Continued)

Name	Country of origin	Products	Website
Outback Power	USA	Charge controllers, inverters	www.outbackpower.com
Xantrax	USA	Charge controllers, inverters	www.xantrax.com
Steca	Germany	Charge controllers, inverters, thermal system controllers, lights, low-energy DC fridges	www.steca.com
SMA	Germany	Larger inverters for off-grid and grid-tied systems	www.sma.de
Solon	Switzerland	Inverters, modules	www.solon.com
Victron	Netherlands	Inverter-chargers, battery monitors	www.victronenergy.com
Mastervolt	Netherlands	Inverters, charge controllers	www.mastervolt.com

Lights

Name	Country of origin	Products	Website
Sundaya	International	Lights, charge regulators, system components	www.sundaya.com
Sollatek	UK	Lights, charge regulators	www.sollatek.com
Quingdao Sparkle Electronic	China	Ballast inverters, LED lights	www.sun-sparkle.com
D.Light Designs	India	Modern low cost lighting systems	www.dlightdesign.com
C.Crane	USA	LED lamps	www.ccrane.com

PV pumps and refrigerators

Name	Country of origin	Products	Website
Lorentz	Germany	Specialized solar pumping solutions	www.lorentz.de
Grundfos	Denmark	Specialized solar pumping solutions	ww.grundfos.com
Sun Pumps Inc	USA	Specialized solar pumping systems	www.sunpumps.com

Refrigerators

Name	Country of origin	Products	Website
SunDanzer	USA	Energy efficient refrigerators and freezers	www.sundanzer.com
Sunfrost	USA	Energy efficient refrigerators and freezers	www.sunfrost.com

Other Products

Name of company	Products	Website
Tracking devices		
Zomeworks	Solar trackers for large arrays	www.zomeworks.com
Wattson	Solar trackers for large arrays	www.wattson.com
solar.trak	Solar trackers for large arrays, solar street lighting	www.solar-trak.de/en
Solar site analysis tools		
Solar Path Finder	Measurement tool to estimate sun path. Also associated software for estimating energy output	www.solarpathfinder.com
Solmetric	Tools for PV site evaluation and energy production estimation	www.solmetric.com
Solar collection and power use meters		
Bogart Engineering	Trimetric charge meter measures voltage, amp collection and daily energy use	www.bogartengineering.com
Kill-a-watt	Meter measures electricity use of appliances	www.p3international.com
DC converters		
Lind Electronics	Specialists in DC-DC converters	www.lindelectronics.com
Zane Inc.	Specialized DC products including dimmers, fan controls, linear power supply	www.zaneinc.com
Solar Converters	Various products for DC power control systems	www.solarconverters.com

Catalogues

The five companies below offer solar products through catalogues and on-line. They have interesting catalogues of products, books and specialized design services. There are many more companies like these around the world.

Name	Country of origin	Products	Website
Real Goods	USA	Large line of solar and off-grid products	www.realgoods.com
Alternative Energy Engineering	USA	Large line of solar products Dealer training Downloadable catalogue	www.aeesolar.com
Phaesun	Germany	Large line of solar and off-grid products Downloadable catalogue	www.phaesun.com
Wind and Sun, Ltd	UK	Large line of solar and off-grid products, excellent printed catalogue	www.windandsun.co.uk
Backwoods Solar (USA)	USA	Specializes in independent electric power systems for remote homes. Provides a useful planning guide and catalogue.	www.backwoodssolar.com

Appendix 1

Energy, Power and Efficiency

The following section is a review of the energy concepts that must be understood to install and design solar energy systems.

Energy

Energy is referred to as the ability to do work. For example, energy is required to boil tea, to move a vehicle between two points or to make a radio work. When boiling tea on a woodstove, the energy source is chemical energy stored in firewood. When driving a car, the source is chemical energy stored in petrol. When operating a radio, the energy source is chemical energy stored in dry cells.

Energy is measured in units called joules, J, or in watt-hours (see below). Because one joule is such a small amount of energy, words that name large numbers of joules are commonly used. One kilojoule, kJ, is equal to a thousand joules, and one megajoule, MJ, is equal to a million joules. Charcoal, for example, contains about 32kJ of energy (or 32,000J) per gram and petrol contains about 45kJ per gram. During the course of a clear day at the Equator, about 23MJ of solar radiation energy falls upon an area of one square metre.

Watt-hours (Wh) are a convenient way of measuring electrical energy. 1 watt-hour is equal to a constant 1 watt of power supplied over 1 hour (3600 seconds). If a bulb is rated at 40 watts, in 1 hour it will use 40Wh, and in 6 hours it will use 240Wh of energy. Electric power companies measure the amount of energy supplied to customers in kilowatt-hours, kWh (or thousands of watt-hours). In this book, energy is always referred to in watt or kilowatt-hours. Note that one kilowatt-hour is equal to 3.6 megajoules.

Power

Power is the rate at which energy is supplied or consumed (or energy per unit of time). Energy can be supplied or consumed at a high rate or at a low rate. For example, it takes roughly the same amount of energy to travel 10km walking as it does to travel 10km running. The difference is that, when running, more energy is being used per unit of time than walking. Similarly, the amount of energy required to boil a pot of water is constant; the time it takes to boil the water depends on the power or the rate at which the energy is supplied. More power is required to boil a pot of water in 2 minutes than is required to boil the same pot in 10 minutes.

Power is measured in watts (W). One watt is equal to one joule supplied per second. As in the case above, large amounts of power are given the name kilowatts, kW (thousands of watts), and megawatts, MW (millions of watts). As an example, an incandescent light bulb might use 40W, while a radio uses about 5W, and an electric cooker might use 2000W. A human being riding a bicycle produces about 200W of power, while a typical automobile engine produce about 25kW. On a clear day, solar power arrives upon a flat surface at a rate of about 1000W (1kW) per square metre.

Efficiency

Efficiency is the ratio of output energy to input energy expressed as a percentage. Mathematically, it is expressed as follows:

$$\text{Efficiency (\%)} = \text{output energy/input energy} \times 100$$

Energy-efficient devices use less energy to perform a given task than energy-wasting ones. For example, some types of stoves use less fuel to cook than others. Similarly, some types of cars use petrol more efficiently than others. Fluorescent lamps consume less energy than incandescent lamps to produce the same amount of light.

Energy Conversions

watt-hours × 1000 = kilowatt-hours

kilowatt-hours × 1000 = megawatt-hours

megajoules/3.6 = peak sun hours or kilowatt-hours

kilowatt-hours × 3.6 = megajoules

langleys × 0.0116 = peak sun hours or kilowatt-hours

langleys × 0.0418 = megajoules

watt-hours × system voltage = amp-hours

Power Conversions

watts × 746 = horsepower

watts × 1000 = kilowatt

kilowatts × 1000 = megawatts

Appendix 2

Basic Extra-Low-Voltage Direct Current (DC) Electricity

This section reviews some terms used to describe basic electrical principles. If you are just beginning to learn about electricity, you should check a secondary-school text for a good introduction to the subject. Books on how to wire houses are a good introduction to the subject of 110/230V AC electricity.

Electricity is power provided by the flow of very small charged particles called electrons through metal wires. Because electrons are so small, it takes millions of them moving together in the same direction to develop a detectable electric current. Wires carrying electricity do not appear any different from wires not carrying electricity (although they may get a bit hot), so electricity is invisible to the human eye when travelling through wires.

Conductors and Insulators

Not all substances can carry electricity. Those that can carry electricity are called conductors and those that cannot are called insulators. Metals such as copper and aluminium are good conductors of electricity, as are salty liquid solutions called electrolytes. Wood, plastic and rubber cannot carry electricity and thus are called insulators. Note that wire cables are wrapped with plastic insulation to prevent the electricity from deviating from its pathway.

The Flow of Electric Current

Although wires are actually very different from water pipes, electricity flowing through wire can be compared to water flowing through pipes. When electricity is flowing, there is said to be electric current.

Current (I) is the rate of flow of electrons through the wires. It is measured in amperes (called amps, A) which is a measure of the number of electrons passing through a given length of wire. This is similar to the rate of flow of water though pipes (i.e. litres per second). Current flowing in one direction is called direct current (DC), while current which changes direction of flow is called alternating current (AC). In a 12-volt DC system, a 12 watt lamp draws about 1 ampere of current.

Voltage, or potential difference, is the difference in potential energy between the ends of a conductor (e.g. a wire) that governs the rate of flow of current through it. Voltage is measured in volts (V). In basic terms, it is the amount of energy each electron has to move about and is similar to the pressure pushing water through a pipe. Grid electricity is supplied at 240 volts AC, while the electricity from automotive batteries is at about 12 volts DC.

Circuit

A pathway which electricity flows through (e.g. the wires, batteries, lamps, switches, etc.) is called a circuit. Current flows from a source of electricity (a battery, generator or solar cell) through wires to loads (lamps, motors, electric coils) and back again. When there is an uninterrupted pathway for electricity to flow, the circuit is said to be closed. When there is a point where electricity cannot pass (i.e. switch turned off), the circuit is said to be open. Thus, when you turn on a light, you close the circuit, and when you turn off a light, you open the circuit. Current cannot flow through an open circuit. Circuit diagrams are pictures of electric circuits with special symbols for switches, batteries, resistive loads, diodes and other electric equipment that help electricians to understand and plan circuits.

Series and Parallel Circuits

When a number of electrical components are wired up end to end in a continuous chain, they are joined in series. If lamps are joined in series and one of them fails, then the circuit will be broken and all the lamps will fail. If batteries or solar cells are joined in series, the voltage increases according to the number of units joined. For example, three 1.5 volt dry cells joined in series will produce a voltage of 4.5 volts.

When components are wired so that one path can be broken without affecting the flow of electricity through the others in the circuit, the components are said to be wired in parallel. The lamps in a mains-wired house are in parallel, and you can turn one off without turning off the rest of the lights in the circuit. When batteries or solar cells are wired in parallel, the available current increases but the voltage stays the same. For example, if the above three 1.5 volt dry cells were wired in parallel, the voltage would remain at 1.5 volts, but the amount of current available would increase.

Basic Electric Laws

In all basic electrical work, the understanding of two formulae is required. Once they are understood, most electrical problems encountered in low-voltage systems can be easily solved. These formulae are simple:

Power Law: watts (W) = volts (V) × amps (A)

Ohm's Law: volts (V) = amps (A) × ohms (Ω)

Power Law

Power (P) is the amount of work the electricity is doing at a given instant. It is measured in watts. The power rating in watts of a light fixture, for example, is a measure of the power it will consume to produce light. Power is calculated by multiplying the voltage (V) by current (I):

Power (P) = Voltage (V) × Current (I) or watts = volts × amps

Examples

1 A globe lamp is connected to a 12-volt battery. When it is turned on, 3 amps of current are flowing through the wire. What is the power of the lamp?
2 A 24-watt DC globe lamp is connected in a 24-volt DC system. When the globe is turned on, what current will be flowing?

Solutions

1 Example 1 is asking for the power of the globe lamp.

Power (P) = Voltage (V) × Current (I) = 12 volts × 3 amps = 36 watts

2 Example 2 is asking for the current flowing through the 24-volt wire.

Current (I) = Power (P) ÷ Volts (V) = 24 watts ÷ 24 volts = 1 amp

Ohm's Law

Resistance (R) is the property of a conductor (i.e. wire or appliance) which opposes the flow of current through it and converts electrical energy into heat. It determines the amount of current that can flow for a certain voltage. Resistance is measured in units called ohms, which are given the symbol Ω.

The formula that relates these three electrical measures is called Ohm's Law:

Voltage (V) = Current (I) × Resistance (R) or volts = amps × ohms

Glossary

alternating current (AC): electric current in which the direction of flow changes at frequent, regular intervals.
amorphous silicon: a type of thin film PV silicon cell having no crystalline structure.
ampere (amp) (A): unit of electric current which measures the flow of electrons per unit time.
ampere hour (amp-hour) (Ah): a measure of total charge commonly used to indicate energy capacity of batteries. 1 amp-hour is equal to the quantity of charge in the flow of 1 ampere over 1 hour.
annual mean daily insolation: the average solar energy per square metre available per day over the whole year.
array: an assembly of several modules on a support structure together with associated wiring.
ballast inverter: a device which converts low-voltage direct current to the type of high-voltage alternating current required by fluorescent lamps.
battery capacity: the total number of amp-hours that can be removed from a fully charged battery or cell at a specified discharge rate.
blocking diode: a solid state electrical device placed in circuit between the module and the battery to prevent discharge of the battery when the voltage of the battery is higher than that of the module (i.e. at night).
bypass diode: a solid state electrical device installed in parallel with modules of an array which allows current to bypass a shaded or damaged module.
cell (battery): the smallest unit or section of a battery that can store electrical energy and is capable of providing a current to an external load.
cell (photovoltaic): see solar cell.
charge controller: a device which protects the battery, load and array from voltage fluctuations, alerts the users to system problems and performs other management functions in a PV system.
charge current: electric current supplied to and stored in a battery.
circuit: a system of conductors (i.e. wires and appliances) capable of providing a closed path for electric current.
circuit diagram: drawing used by electricians to represent electric circuits.
connector strip: insulated screw-down wire clamp used to fasten wires together in electric systems.
converter: a device that converts a DC voltage source to a higher or lower DC voltage.
crystalline silicon: a type of PV cell made from a single crystal or polycrystalline slice of silicon.
current (amps, amperes) (A): the rate of flow of electrons through a circuit.
cycle: one discharge and charge period of a battery.
cycle life: the number of cycles a battery is expected to last before being reduced to 80 per cent of its rated capacity.
days of storage: the number of consecutive days a stand-alone system will meet a defined load without solar energy input.
deep-discharge battery: a type of battery that is not damaged when a large portion of its energy capacity is repeatedly removed.
depth of discharge (DoD): a measure in percentage of the amount of energy removed from a battery during a cycle.
design month: the month which has the lowest mean daily insolation value, around which many stand-alone systems are planned.

diffuse radiation: solar radiation that reaches the earth indirectly due to reflection and scattering.

direct current (DC): electric current flowing in one direction.

direct radiation: radiation coming in a beam from the sun which can be focused.

discharge: the removal of electric energy from a battery.

efficacy: term referring to the efficiency by which lamps convert electricity to visible radiation, measured in lumens per watt.

efficiency: the ratio of output power (or energy) to input power (or energy) expressed as a percentage.

electrolyte: a conducting medium in which the flow of electric current takes place by migration of ions. Lead-acid batteries use a sulphuric acid electrolyte.

equalizing charge: a charge well above the normal 'full' charge of a battery which causes the electrolyte inside the cells to bubble and get mixed up.

extra-low voltage: the IEC defines extra-low voltage as not more than either 50V AC (70V RMS peak), or ripple-free 120V DC, in relation to earth/ground.

fuse: a device which protects circuits and appliances from damage caused by short circuits and current overloads by cutting off the supply.

global radiation: term which refers to the combined diffuse and direct solar radiation incident on a surface.

hydrometer: a tool which indicates the state of charge of lead-acid batteries by measuring the thickness of the acid inside its cells.

insolation: incident solar radiation. A measure of the solar energy incident on a given area over a specified period of time. Usually expressed in kilowatt-hours per square metre per day or indicated in peak sun hours.

inverter: a solid state device which changes a DC input current into an AC output current.

irradiance: the solar radiation incident on a surface per unit time. Expressed in watts or kilowatts per square metre.

I-V curve: the plot of current versus voltage characteristics of a solar cell, module or array. I-V curves are used to compare various solar cell modules and to determine their performance at various levels of insolation, loads and temperatures.

kilowatt (kW): 1000 watts. Standard method of measuring electrical power.

kilowatt hour (kWh): energy equivalent to 1000 watts delivered over the period of 1 hour. Standard method of measuring electrical energy.

langley (L): unit of solar insolation ($1L = 85.93\ kWh/m^2$).

light-emitting diode (LED): a lighting device made from a diode which illuminates when current is flowing through it. Also, panel indicator on charge controllers or electrical devices.

load: the set of equipment or appliances that use the electrical power from the generating source, battery or module.

low voltage: the IEC defines low voltage as any voltage in the range 50–1000V AC or 120–1500V DC.

low-voltage cut-out or disconnect (LVD): a feature of some charge controllers that cuts off power to the load when the battery reaches a low state of charge.

maximum power point (MPP): the specific point or voltage where, under given conditions, the module produces the greatest power. This can be identified on an I-V curve.

monthly mean daily insolation: the average solar energy per square metre available per day of a given month.

ohm (Ω): a unit of electrical resistance.

open circuit voltage (Voc): the maximum possible voltage across a solar module or array. Open circuit voltage occurs in sunlight when no current is flowing.

overcharging: leaving batteries on charge after they have reached their full (100 per cent) state of charge.

peak power (Wp): the amount of power a solar cell module can be expected to deliver at noon on a sunny day (i.e. at Standard Test Conditions or STC) when it is facing directly towards the sun.

peak sun hours: the number of hours per day during which solar irradiance averages 1000W/m^2 at the site. A site that receives 6 peak sun hours a day receives the same amount of energy that would have been received if the sun had shone for 6 hours at an irradiance of 1000W/m^2.

photovoltaic (PV) device: a device which converts light energy into electric energy.

potential difference (voltage) (V): the difference in potential energy between the poles of a conductor that governs the rate of flow of current. Measured in volts (V).

power conditioning unit (PCU): electrical equipment used to convert DC power from a PV array or battery into a form suitable for standard AC loads (240 or 110V AC). PCUs are used to operate high-voltage appliances such as videos and refrigerators.

resistance: the property of a conductor (i.e. a wire or appliance) which opposes the flow of current through it and converts electrical energy into heat. Resistance has the symbol R and is measured in ohms (Ω).

self-discharge: charge lost from batteries left standing due to reactions within the cells.

shallow discharge batteries: batteries designed to supply high power for a short duration; taking too much energy out of these batteries before recharging them is likely to damage the plates inside (e.g. SLI or automotive batteries).

short-circuit current (Isc): current across the terminals when a solar cell or module in strong sunlight is not connected to a load (measured with ammeter).

silicon: a semiconductor material commonly used to make photovoltaic cells.

solar cell: a specially made semiconductor material (e.g. silicon) which converts light energy into electric energy.

solar cell module: groups of encapsulated solar cells framed in glass or plastic units, usually the smallest unit of solar electric equipment available to the consumer.

solar constant: an unchanging value that refers to radiation arriving from the sun at the edge of the Earth's atmosphere. The accepted value is about 1300 watts per square metre.

solar incident angle: the angle at which the incoming solar beam strikes a surface.

specific gravity: the ratio of the weight of a solution (e.g. battery acid) to an equal volume of water at a specified temperature. Used as an indicator of battery state of charge.

Standard Test Conditions (STC): a set of accepted testing conditions commonly used by manufacturers to compare solar cell modules of different types. The conditions are 1000W/m^2 solar irradiance at 25°C with an air mass of 1.5.

stand-alone solar electric system: a solar electric system that receives all of its energy from solar electric charge, and which is not connected to the grid or any other source of power, also called off-grid.

state of charge (SoC): the amount of charge in a battery expressed as a percentage of its rated charge capacity.

system voltage: the voltage at which the charge controller, lamps and appliances in a system operate, and at which the module(s) and battery are configured.

total daily system energy requirement: the amount of energy required to meet the daily electrical load plus the extra energy required to overcome system energy losses.

tracking: the practice of changing the position (i.e. angle) of the array over the course of the day so that it faces the sun and so harvests a larger amount of solar charge.

trickle charge: a low current charge. When the batteries are fully charged, charge controllers reduce the energy from the module to the battery to a trickle charge so that the batteries are not overcharged, but so that they still get enough current to overcome self discharge.

volt (V): a unit of measurement of the force given to electrons in an electric circuit; see potential difference.
voltage drop: loss of voltage and power due to the resistance of the wire to the flow of electricity in long runs of cable.
watt (W): the internationally accepted measurement of power. 1000 watts is a kilowatt, and 1 million watts is a megawatt.
watt-hour (Wh): a common energy measure arrived at by multiplying the power times the amount of time used. Grid power is ordinarily sold and measured in kilowatt-hours.

Worksheets

The design worksheets on the following pages are for helping to plan small off-grid solar electric systems. Their use is fully explained in Chapter 8. Worksheet 1 is for use in selecting loads and estimating the energy demand of a system. Worksheet 2 is for use in sizing and selecting the solar array. Worksheet 3 is for use in sizing and selecting the battery bank. Worksheet 4 is for use in sizing and selecting the charge controller and inverter. Worksheet 5 is for use in sizing and selecting the cables, fuses and associated components.

Excel versions of the worksheets can be downloaded from www.earthscan.co.uk/expert. The downloadable versions complete all of the calculations automatically. The downloadable worksheets will be improved occasionally and other information may be available periodically at the same site.

Please note that, although useful, these design worksheets only simulate off-grid solar PV systems. They are not perfect, and the successful performance of any solar PV system depends on how the user (i) operates and maintains the system, and (ii) responds to real world weather patterns.

The following shades indicate how cells should be used:

Enter data in this cell
See text for a description of this cell
This cell contains a calculation and is locked

Worksheet 1: Daily System Energy Requirement

Use this worksheet to calculate the energy which must be supplied by the array each day to power all of the lights and appliances in your system

1. Calculate the total daily load energy demand using the table below

Column A Lamp or Appliance list below	Column B Voltage volts	Column C Power watts	Column D Daily Use hours	Column E Daily Energy Use (DC) watt hours	Column F Daily Energy Use (AC) watt hours
DC Appliances					
AC Appliances					

BOX G: Total Daily DC Energy Demand ______ Watt-hours

BOX H: Total Daily AC Energy Demand ______ Wh

2. Estimate system energy losses

Energy is always lost due to inefficiencies in cables modules, batteries, charge controllers and inverters. The extra amount of energy lost must be estimated and added to the daily energy demand.

For DC power multiply Box G by	0.20
For AC power multiply Box H by	0.35

Box I: DC losses ______ Wh

Box J: AC losses ______ Wh

3. Add AC & DC Demand and AC & DC Losses

Box K: Total Daily System Energy Requirement ______ Wh

4. Select System Voltage

Box L: System Voltage 12 volts

5. Calculate Daily System Charge Requirement

Divide Box K by the system voltage in Box L

Box M ______ Ah

Worksheet 2: Sizing and Choosing the Modules

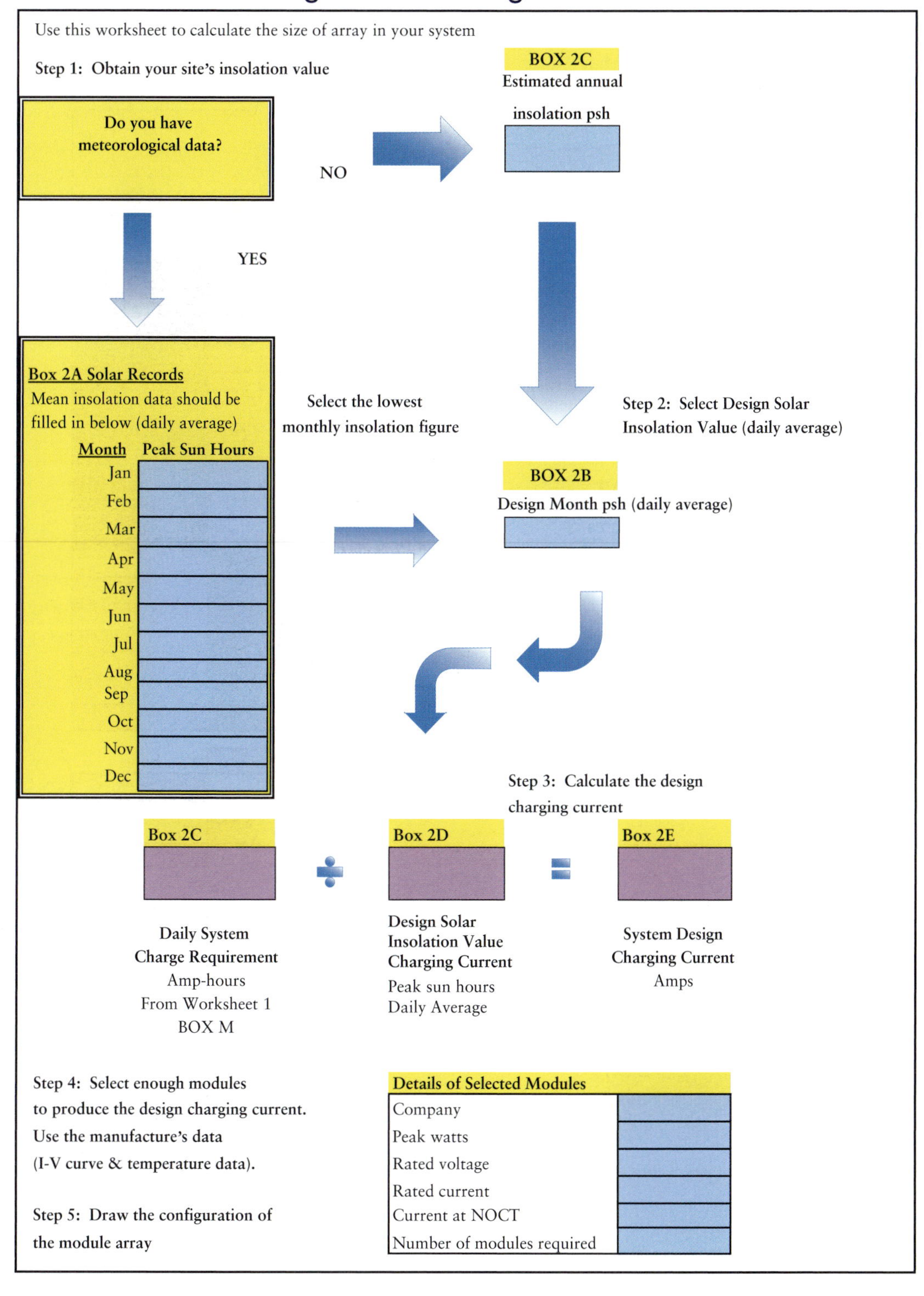

Worksheet 3: Battery Sizing and Selection

Use this worksheet to select your battery and charge controller

Step 1: What type of battery will you use?
Shop around to get an idea of the batteries available (see Table 4.3)

Type of Battery	Voltage	Capacity	Recommended Daily DoD (%)	Recommended Max DoD (%)	Price	Cycle Life @ 25% DoD
Modified SLI (Portable)						
Traction Battery						
Lead Calcium Maintenance Free						
Captive Electrolyte Gel Cells						
Absorbed Glass Mat						
Tubular Plate Batteries						
Other						

Step 2: Calculate the capacity required capacity of your battery in amp hours.

Box 3A	×	Box 3B	÷	Box 3C	=	Box 3D
Daily System Charge Requirement Amp-hours From Worksheet 1 BOX M		**Reserve Days** Minimum 1 Maximum 4		**Maximum DoD** Express as decimal (ex. 30% is 0.30)		**Required System Battery Capacity** Amp-hours

Step 3: Determine the battery configuration

Make a drawing of battery configuration/voltage

Indicate the voltage and parallel/series connections

Step 4: Does battery require special charge control setting?

YES ☐
NO ☐

Step 5: Determine your battery cycle life and maintenance needs.

Maintenance needs		How often
Topping up	YES/NO	Monthly
Cleaning corrosion	YES/NO	Monthly
Equalizing	YES/NO	Quarterly
Check SoC	YES	Weekly

Expected Life Cycle: ______ Cycles
______ (Days)

Details of Selected Battery	
Company	
Model	
Type	
C20 Capacity A	
Individual Battery Voltage	
Battery Bank Voltage	
Number in series	
Number in parallel	
Total number of batteries	
Total battery capacity	
Rated cycle life @ 25% DoD	

Worksheet 4: Charge Controller/Inverter Sizing and Selection

Use this worksheet to size and select your charge controller and inverter

Charge Controller Sizing and Selection

Step 1: Select Charge Controller Size

Array Input Rating			DC Load Output Current Rating		
Box 4A		Box 4B	Box 4C		Box 4D
	× 1.25 =			× 1.25 =	
Maximum Array Short Circuit Current Amps		Array Input Rating Amps	Maximum DC Load		Load Output Rating Amps

Step 2: Select Desired CC Features

Shop around to get an idea of the charge regulators available (see Table [])

Controller Specification		Rating	Protection	YES/NO
Rated Voltage	V	See Box L	Short circuit protection (array)	
Maximum Array ISC Input	A	See Box 4B	Short circuit protection (load)	
Maximum Load Output	A	See Box 4D	Reverse polarity (array)	
Self Consumption	mA		Reverse polarity (load & battery)	
Feature	YES/NO		Lightning protection	
High voltage disconnect			Open circuit battery	
Low voltage disconnect			Sealed battery charge settings	
Temperature compensation				
Load timer				

Displays	YES/NO	Desired Charging Type	YES/NO
Solar charge indicator		Pulse Width Modulation	
SoC indicator (LED)		Equalization	
LCD display		MPPT	
Amp hour meter		Other	

Inverter Sizing and Selection

	YES/NO	
Step 1: Is an Inverter Needed?		Only if 230/110 VAC appliances are used.
Step 2: Inverter/Charger?		Only if a charging source is available

Step 3: Inverter & Charger Rating

Box 4E		Box 4F
	× 1.25	
Maximum AC Load Watts		Inverter Rating Watts

Rated Charge Current

For battery charging (A)

Accepts power from:

Step 4: Select Desired Inverter Features

Details of Desired Inverter	
Continuous Power (W)	
Peak Surge Power (W)	
Input DC Volts	
Output AC Volts	
Wave shape	
Efficiency	
Self-Consumption (A)	
Low Voltage Disconnect (V)	

Worksheet 5: Wiring, Voltage Drop and Fuses

Use this worksheet to size and select your electrical accessories

Step 1: Map the Site and Estimated Cable Runs

Draw a scale map of the site and estimate the distance of all cable runs, including from the modules to the control, the control to battery and the circuits. Note the locations of each lamp, socket, switch, junction box, and fuses. Note the voltage of major circuits.

Step 2: Determine the proper cable sizes to avoid voltage drop

See Chapter 7 for instructions on the use of the table below.

Column A Cable Run (List Circuits)	Column B Distance of Cable (m)	Column C Maximum Current (A)	Column D K Value of wire (ohms/m)	Column E Total Resistance (ohms)	Column F Voltage Drop (V)	Column G Voltage Drop (%)
						%
						%
						%
						%
						%
						%
						%
						%
						%

Make sure all voltage drops are less than 2%!

Step 3: Sizing Fuses and Circuit Breakers

See Chapter 7 for detailed instructions on sizing fuses

- List circuits to be protected in Column A
- Write each circuit's maximum power drawn (W) in Column B
- Calculate each circuits' maximum current (A) in Column C
- Increase the figure by 20% in Column D

This is the minimum size of the fuse required.

Column A Cable Run (List Circuits)	Column B Max Rated Power (W)	Column C Maximum Current (A)	Column D Fuse Size (A)

Step 4. List all electrical accessies required.

- Cables (all sizes & types)
- Conduit
- Switches
- Sockets
- Fuses
- Junction boxes
- Connector strips
- Lightning protection
- Grounding/earthing
- Bolts, screws, nuts, etc.

Use the list of electrical accessories to make sure you have everything you need before going to site!

Index